The
Flightless
Traveller

First published in Great Britain in 2020 by
Greenfinch
An imprint of Quercus Editions Ltd
Carmelite House
50 Victoria Embankment
London EC4Y 0DZ

An Hachette UK company

A CIP catalogue record for this book is available
from the British Library

HB ISBN 978-1-52941-072-3
E-Book ISBN 978-1-52941-073-0

Every effort has been made to contact copyright holders. However, the publishers will be
glad to rectify in future editions any inadvertent omissions brought to their attention.

Quercus Editions Ltd hereby exclude all liability to the extent permitted by law for any
errors or omissions in this book and for any loss, damage or expense (whether direct or
indirect) suffered by a third party relying on any information contained in this book.

10 9 8 7 6 5 4 3 2 1

Design by Nathan Burton
Illustrations by Taku Bannai

Printed and bound in Italy by L.E.G.O SpA

Papers used by Greenfinch are from well-managed
forests and other responsible sources.

The Flightless Traveller

50 modern adventures
by land, river and sea

Emma Gregg

greenfinch

Contents

Fly less and travel better 6

HOW TO TRAVEL 8

Travelling the world by rail, river and sea 10
What counts as sustainable tourism? 12
Flight-free FAQs 16
The greenest ways to travel 24
The bigger picture 30

THE TRIPS

Green City Breaks

Get to know Bristol by bike, boat and hot-air balloon 34
See Paris in a new light 38
Decamp to Den Bosch, a hidden highlight of the Low Countries 42
Cycle to your heart's content in Copenhagen 46
Wander around Freiburg, a green city in the Black Forest 50
Soak up the creative vibes in Dublin and Galway 54
Flit between Berlin and Vienna by rail 58
Fall in love with the soul of Iberia in Madrid and Lisbon 62
Stimulate your senses in Tangier, Casablanca and Marrakech 66
Dose up on architecture in Helsinki, Tallinn and St Petersburg 70

Blue Sea Thinking

Tour Scotland's western isles, from Arran to Skye 76
Recreate your childhood summers in Scilly 82
Discover the hidden beaches of Finistère 86
Escape to Formentera's endless sands 90
Ride the waves in Ericeira 92
Island-hop from the Côte d'Azur to Calabria 94
Unwind on a Mediterranean gulet 100
Kayak your way along the Dalmatian Coast 104
Choose your island in the Stockholm Archipelago 108
Relax on Essaouira's Plage Tagharte 112

Railway Stories

Steam across the Harz Mountains 116
Spiral through the Alps on the Bernina Express 118
Arrive in style in Arlberg, the heart of the Austrian highlands 122
Explore the Flåm Valley by train 126
Travel like a tsar from Sofia to the Black Sea 130

Adventures in the Slow Lane

Backpack along the Wales Coast Path 136
Put your best foot forward on the West Highland Way 142
Follow the Travellers' trails of County Wicklow 146
Bird-watch by bike, from Denmark to Belgium 150
Mess about in an electric canal boat in Alsace 154
Go cross-country skiing in the French Alps 156
Stride along El Camino de Santiago, the pilgrims' route 160
Gallivant through Bavaria on the King Ludwig Way 164
Cruise along the River Danube 168
Stroll through Slow Food country in Piedmont 172
Ride an electric bike through the Tuscan hills 176
Explore the Skåneleden on foot 178
Wend your way through southern Transylvania by bike 182
Paddle the River Soča 186
Feel on top of the world in the Atlas Mountains 188

Epic Voyages

Cross Europe by bus, from Paris to Kyiv 194
Eight time zones without flying: London to Singapore 198
Take the Trans-Siberian Railway to Tokyo 204
The ultimate overland adventure: experience Africa from top to toe 208
Harness the elements on a transatlantic yacht rally 214
Board a tall ship for a low-carbon cruise on the high seas 218
Voyage from Italy to Australia by cargo ship 222
Head north on an expedition to Iceland and Greenland 228
Go with the floe: Barcelona to Antarctica 234
Footloose and flight-free: circle the world by land and sea 240

THE GREEN TRAVELLER'S DIRECTORY 246

FLY LESS AND TRAVEL BETTER

'We cannot be radical enough in dealing with the issues that face us at the moment. The question is what is practically possible.'
Sir David Attenborough, broadcaster and naturalist

'What you do makes a difference, and you have to decide what kind of difference you want to make.'
Dr Jane Goodall, DBE, primatologist and conservationist

We live in a world where we can dream up a plan to visit a city half the world away one morning, and be there the next. But are we travelling wisely and well?

Choosing to travel by land, river and sea instead of flying provides an opportunity to travel better. It allows us to reconnect with the planet: its contours, textures, intricate details and huge skies.

Some of the best journeys we'll ever take are those that appeal to the problem-solver in us: there's a route to plan and a budget to juggle. They'll involve new experiences: travelling as a passenger in a cargo ship, perhaps, kayaking along an Eastern European river, spending a few days as the houseguest of a family in remotest rural Canada, or discovering flavours we've never tasted before. Often, they'll encourage us to explore the path less travelled, getting to know people we'd never normally encounter and finding answers to questions we'd never think to ask.

Other flight-free trips simply extend the time it takes to reach our destination; adding interesting visits, meals and overnight stops to make the journey an absolute pleasure rather than a chore.

Voyages by land, river and sea needn't be complicated. They can be blissfully simple. Picture yourself throwing a few things in a pannier, jumping on your bike and setting off.

What other benefits can we enjoy when we fly less? Rather than hurtling between time zones, altitudes and climates, risking jet lag and culture shock, we can proceed more gradually, letting our bodies and souls adjust naturally to the changes we witness along the way.

Importantly, moderating our flying habits – eliminating inessential flights and flying far less frequently – is a highly effective means of reducing our carbon footprint.

So let's stop taking travel for granted, and turn it back into an art.

Travel safely, travel well, and enjoy the journey.

HOW TO TRAVEL

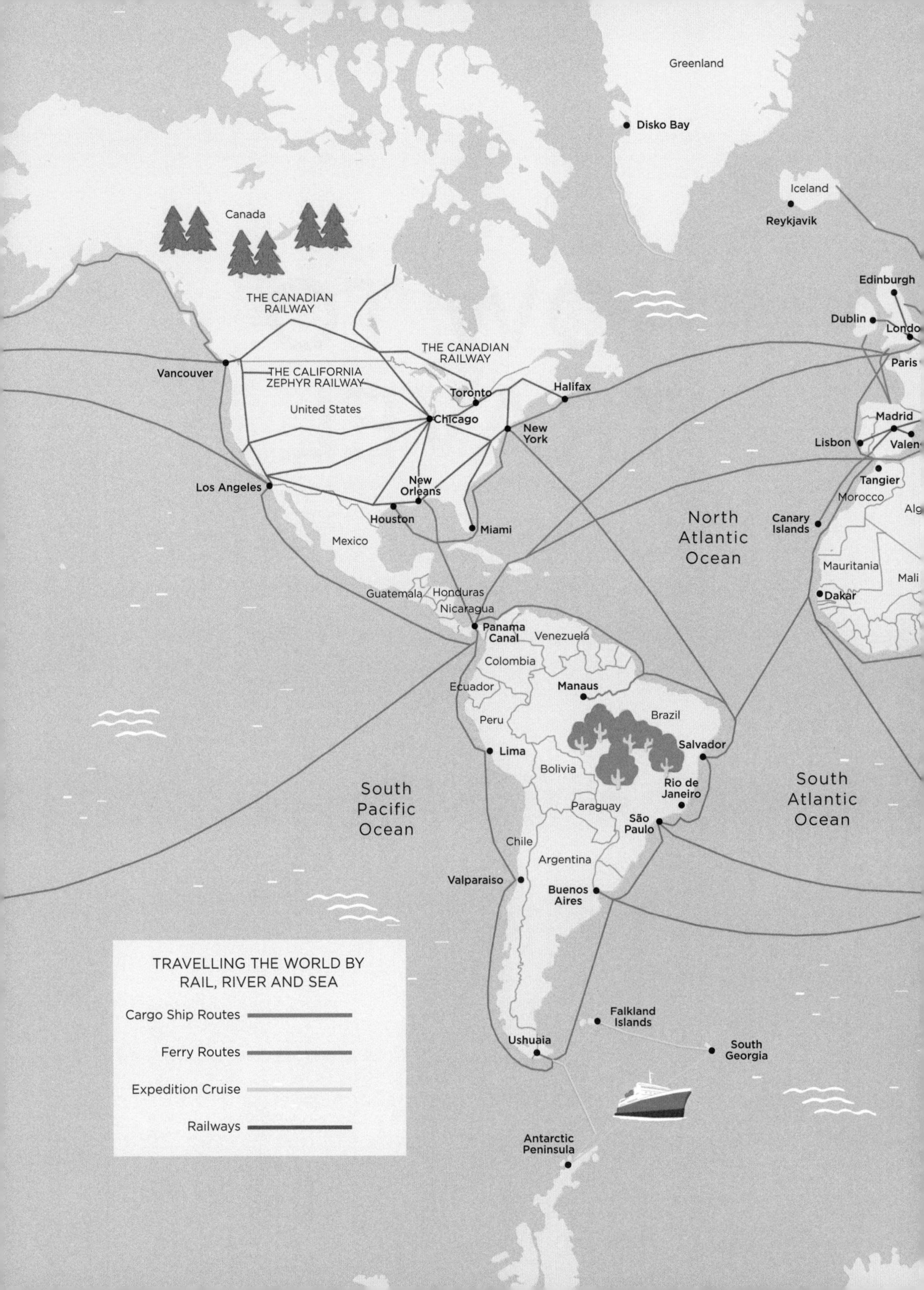

TRAVELLING THE WORLD BY
RAIL, RIVER AND SEA

Cargo Ship Routes

Ferry Routes

Expedition Cruise

Railways

Greenland

Disko Bay

Iceland

Reykjavik

Edinburgh

Dublin

Londo

Paris

Canada

THE CANADIAN
RAILWAY

THE CANADIAN
RAILWAY

Vancouver

THE CALIFORNIA
ZEPHYR RAILWAY

Toronto

Halifax

United States

Chicago

New
York

Madrid

Lisbon

Valen

Los Angeles

New
Orleans

Houston

Miami

Tangier

Morocco

Alg

Canary
Islands

Mexico

North
Atlantic
Ocean

Mauritania

Mali

Guatemala

Honduras

Nicaragua

Dakar

Panama
Canal

Venezuela

Colombia

Ecuador

Manaus

Peru

Brazil

Lima

Salvador

Bolivia

Rio de
Janeiro

South
Pacific
Ocean

Paraguay

São
Paulo

South
Atlantic
Ocean

Valparaiso

Chile

Argentina

Buenos
Aires

Falkland
Islands

Ushuaia

South
Georgia

Antarctic
Peninsula

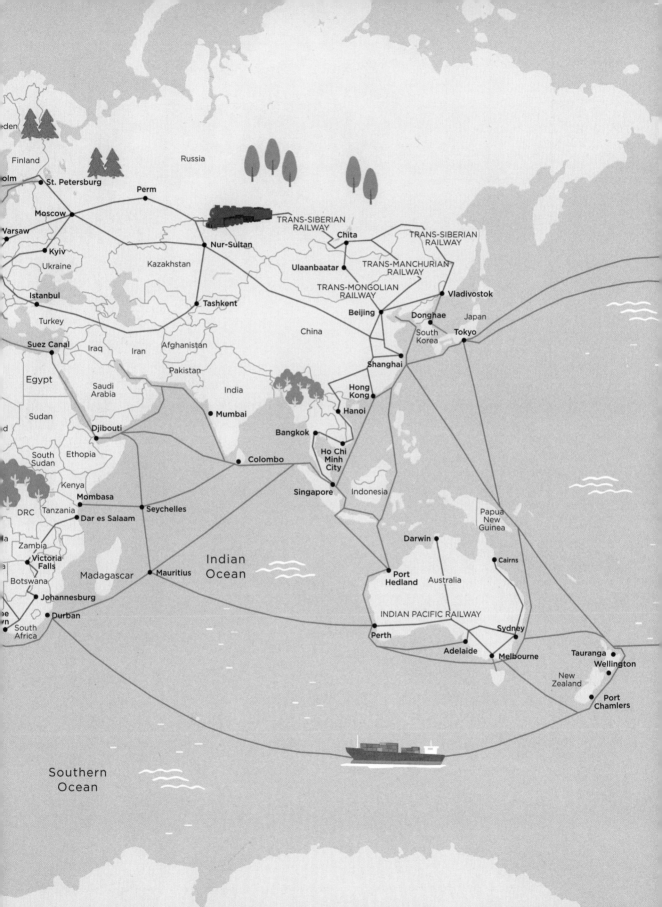

WHAT COUNTS AS SUSTAINABLE TOURISM?

Think back over all your travel experiences so far. How would you rank them?

The trips that were relaxing, thrilling, rejuvenating, enlightening, fun – or a combination of all of the above – will probably be up near the top. But chances are, the memories that shine brightest will be those moments when, whether you realized it at the time or not, you made a difference. That occasion when you swapped stories with a local family over an unforgettable meal you helped prepare. The time you snorkelled over a rescued coral reef that's now pristine, against the odds, thanks to a local marine conservation programme funded by tourism. Or the day that you shared one of your gentle passions with your kids – bird-watching, say, or listening to up-and-coming musicians – and saw delight in their eyes.

The truth is, most of us are responsible, sustainable travellers at heart. We love the thought that our trips away from home do no real harm. If they have some kind of positive effect on the places we visit and the people we meet, so much the better.

Tourism has the power to do this, and more. It can be an overwhelming force for good. It can transfer wealth to those in need – from the developed to the developing world, for example, or from urban to rural communities – by supporting entrepreneurs and generating worthwhile jobs. It can save priceless cultural treasures from being shattered, forgotten or lost. It can help conserve forests, wetlands, reefs and our planet's remarkable biodiversity, both for their own sake, and for our best chance of survival in the face of climate change.

Sustainable, ethical, green: is there a difference?

When planning a trip that's kind to the planet and considerate towards its communities, there are several possible approaches, each with a slightly different focus.

Sustainable tourism has long-term positive effects. It operates within realistic limits, so there's no danger of environmental damage through too many tourists taking part. It doesn't skew local education programmes or economies, either – encouraging people to skip schooling and abandon farms, for example, in order to tout for guiding work instead.

Ecotourism has helped Costa Rica restore vast swathes of its cloud forests and rainforests. This enterprising nation is now a biodiversity hotspot once more.

Ethical tourism weighs up what's right and what's wrong. There are many factors to consider. Does your trip have an acceptably low environmental impact? Do you feel it's fair to the host community, including children, women and animals? Your answers will, to some extent, be subjective.

Green tourism minimizes carbon emissions and maximizes air quality, habitat protection and biodiversity. In this era of climate crisis and mass extinction, it's critically important. Crucially, its success depends on communities being actively involved, and receiving tangible benefits such as jobs, resources and security. People who live close to dangerous wild animals, for example, require appropriate support.

Community-friendly tourism focuses on making destinations better places to live, partly by ensuring that local people have a fair stake in the many benefits that tourism can bring. Sometimes, it might enable tourists and hosts to interact in an open, authentic and enjoyable way, through homestays, crafts, culinary experiences and tours. It could also celebrate and conserve local culture and heritage in a manner that's dignified, respectful and beneficial to locals and visitors alike.

Responsible tourism binds all these principles together. It involves making ethical decisions that result in sustainable, green, community-friendly travel experiences.

Independent tourism is all about planning your own itinerary and booking transport, accommodation, meals, activities and guides that suit you best.

Cultural tourism teaches you about customs, lifestyles, history and heritage, usually through direct contact with local people.

Ecotourism immerses you in nature, often in the company of an expert local guide.

Not all independent, cultural or ecotourism trips are responsible and sustainable, but the best examples most definitely are. They're often the most enjoyable travel experiences you'll ever have, as well.

Calculating the true cost of tourism

In recent years, those of us who live in the world's wealthiest societies have developed a blind spot. We're well aware that many modes of transport – particularly planes, lorries and ships – create high carbon emissions, atmospheric pollution, noise and other disturbances. Tourism, which relies heavily on air and road transport, is complicit in this. But we often accept the damage as unfortunate but inevitable, and feel powerless to act.

Unfortunately, we can no longer look the other way. In 2019, the World Tourism Organization predicted that by 2030, the number of flights tourists make each year will be 592 per cent higher than it was in 2005. In our eagerness to see more of the world, we're in danger of ruining it.

As a result of our collective decisions, the air everyone breathes is compromised. New runways, roads and ports destroy precious habitats. Wild creatures are injured and killed by collisions with aircraft, vehicles and tankers. And with more and more of us travelling, another problem has arisen: over-tourism. Some of our favourite destinations can barely cope.

Applying our imagination, one trip at a time

So is the answer to stop moving about, and simply reconnect with the beauty and fascination of our home neighbourhood?

In some ways, yes. Dialling down our appetite for travel and appreciating everything we have on our own doorstep that little bit more could have a hugely positive effect.

But in other ways, no. If we stopped travelling, the long-term impact on intercultural understanding, conservation and the global balance of wealth could be catastrophic. And if we simply eliminated tourists' most harmful modes of transport – cruise ships and aircraft – island and wilderness communities would end up chronically isolated. Some of these people are the guardians of our most precious natural resources. If we lose them, we lose everything.

Plus, of course, if we decided we weren't going to travel any more, life wouldn't be as exciting and rewarding.

As a way forward, let's plan our travels thoughtfully, carefully considering our mode of transport before each trip. In carbon emission terms, some flights are not as bad as others: large passenger planes have lower emissions per kilometre (or mile) over journeys of 5,000–6,500km (3,000–4,000 miles) – London to Addis Ababa or New York to Lima, say – than over shorter distances (because of the high carbon cost of landing and take-off) or longer distances (that require a heavy fuel load).

Furthermore, some justifications for flying are stronger than others: there's a good case to be made for making occasional visits to far-flung relatives and dearest friends, for example, or trips to remote conservation or development projects that rely on international visitors for their survival. In such instances, let's stay as long, and do as much, as we can, in order to make every flight count.

As for inessential flights: let's show them the door by giving due consideration to less harmful modes of transport. Let's apply our imagination. You never know where the adventure might lead.

Even if we're on top of the science, discussing the climate crisis can feel uncomfortable. Many of us, reluctant to appear judgmental, would hate to provoke a case of *flygskam* (flight shame). Who wants to pour cold water on somebody else's holiday plans?

One way to gain confidence is to connect with people who approach flightlessness with an open mind. Campaigns such as Flight Free UK, Stay Grounded and We Stay on the Ground encourage travellers to share their experiences, upsides, downsides and all.

First-hand tales of fabulous journeys can be empowering. You'll soon be visualizing your next flight-free trip and encouraging others to do the same. Since airlines and policy-makers respond to public opinion and demand, sharing the joys of flying less could make a measurably positive difference to the environment within our lifetimes.

FLIGHT-FREE FAQS

Our love affair with flying crept up on us, and many of us are having second thoughts. When was it, for you, that overseas trips by plane changed from a longed-for treat to a commonplace event? When did quick city breaks on budget airlines get sexy, letting your desire for fewer, longer holidays fade away? When did you get into the habit of scanning the internet for cheap flights, and choosing your next summer holiday destination accordingly? And when did you start wondering whether these changes were actually for the best?

Is it time to rethink? Whether you're old enough to remember a time when everyone dressed up for a flight or young enough for the era of low-cost air travel to be all you've ever known, you're sure to have plenty of questions.

Is flight-free travel really better for the environment?

Confronted with overwhelming evidence that we must take urgent action to reduce pollution and halt climate change, it's become hard to justify inessential flights. Flying is the single most climate-polluting activity an ordinary person can do.

Nitrogen oxide and carbon dioxide emissions from aircraft and airport operations contribute significantly to global heating, and this contribution is growing. According to the World Tourism Organization, journeys to tourist destinations will produce over 1,750 megatonnes of greenhouse gases each year by 2030. While only 32 per cent of tourists will fly, their flights will account for the lion's share of the emissions, at around 56 per cent.

It can be hard for individual air passengers to get a handle on what their travel decisions mean in practice, but carbon calculators can help. They can show us that each time we fly, our personal carbon footprint increases by a significant percentage.

Flying is not the only culprit. Other holiday activities can be harmful to the environment: driving large cars with poor fuel efficiency or travelling on spacious, luxurious, oil-burning cruise ships, for example, has an overwhelmingly negative impact. It's no wonder that people are beginning to turn away from these in favour of choices that align better with their life goals and values.

A return flight taking 2.5 hours each way (London to Madrid, say) adds around 0.4 tonnes of carbon dioxide to your footprint: more than six times as much as you'd clock up by train. This is a large chunk of the 2.1 tonnes per year that individuals should aim for if we're to share emissions fairly and keep the global temperature rise to below 2°C, as set out in the Paris Agreement. The impact of those 0.4 tonnes is similar to failing to recycle for two years or, if you normally eat a plant-based diet, switching to meat for six months.

WHY DO CARBON CALCULATORS GIVE SUCH DIFFERING RESULTS?

The simplest calculators estimate the average fuel burn on a 'tank to wheel' basis, ignoring whether or not the fuel was renewably sourced. Others take into account fuel efficiency and passenger capacity, and take a 'well to wheel' perspective, factoring in the energy used in producing the fuel.

Is flight-free travel better for our health and wellbeing?

No mode of transport is problem-free, but there are medical risks involved in flying that not all options share. Since airline passengers sit close together and share washrooms, there's a high danger of direct transmission of viruses. Wider dispersal of airborne microorganisms poses less of a problem, since scientific studies have shown that, contrary to popular belief, the air in planes is at least as clean as in the average office, with cabin circulation systems filtering out between 94 and 99.9 per cent of airborne microbes and providing a total changeover of air every few minutes. Reassuring as this may be, it's not all good news. Cabin air causes dehydration, resulting in dry skin and flagging energy.

Other potential problems such as deep vein thrombosis, hypertension and anxiety can be mitigated through physical and mental exercises, but jet lag is trickier to tackle unless you're used to calibrating your body clock to different time zones. Aircraft noise, which can exceed 140 decibels, has adverse effects on everyone on board, and on people and wildlife in the vicinity.

Is flight-free travel practical, timing-wise?

There are plenty of good reasons to fly less and use other modes of transport instead. But given that most of us are bound by time constraints, is travelling without flying actually realistic?

Every day we devote to trips away from home is precious. It's understandable, therefore, that flying might feel like an essential part of a 'proper' holiday, despite its downsides. On paper, it seems to save time.

The thing is, it consumes time, too. It's not just the time in the air: it's also the preamble, with all its frustrations. Could those hours be better spent? Most of us live far closer to railway and bus stations than to airports, and boarding a train or bus takes a fraction of the time that airport procedures require.

The closer the destination, the harder it is to argue that flying is faster, particularly on short routes where there's a high-speed train that will do the job. Over longer distances, flying does gain you time. But consider what you lose – the opportunity to travel through different regions, for example, and the thrill of adventure. It's well worth rethinking your annual plans to accommodate fewer, lengthier trips, sacrificing a little time in each destination but making the most of each journey.

Travelling under your own steam can deliver the heady buzz of freedom and a raft of rich experiences, from the enjoyable responsibility of planning your own route to the pleasure of serendipitous encounters and discoveries – that friendly fellow cyclist you meet on the trail, for example, or the cosy little pub you come across by chance on your overnight stop.

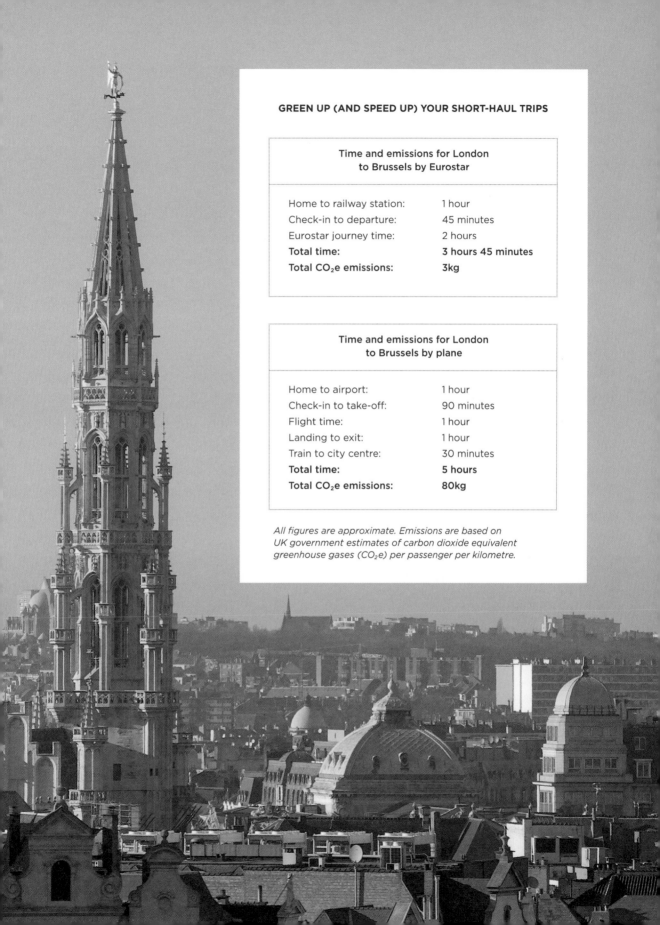

GREEN UP (AND SPEED UP) YOUR SHORT-HAUL TRIPS

Time and emissions for London to Brussels by Eurostar	
Home to railway station:	1 hour
Check-in to departure:	45 minutes
Eurostar journey time:	2 hours
Total time:	**3 hours 45 minutes**
Total CO$_2$e emissions:	**3kg**

Time and emissions for London to Brussels by plane	
Home to airport:	1 hour
Check-in to take-off:	90 minutes
Flight time:	1 hour
Landing to exit:	1 hour
Train to city centre:	30 minutes
Total time:	**5 hours**
Total CO$_2$e emissions:	**80kg**

All figures are approximate. Emissions are based on UK government estimates of carbon dioxide equivalent greenhouse gases (CO$_2$e) per passenger per kilometre.

Is flight-free travel good value?

If you're satisfied that flying is bad for the planet and you're willing to spend time on healthier alternatives, what's holding you back?

If it is a concern about cost, you're not alone. Ever since the 1960s, when package holidays and charter flights were launched, we've increased our appetite for affordable international travel. In the 1990s, budget airlines pushed things further by suggesting we all deserved to take several international flights a year, and we'd be crazy to spend more than the bare minimum on them. Meanwhile, train and bus fares seemed to go up and up, particularly for those unwilling, or unable, to book special offers in advance.

By avoiding flying, you avoid overpriced airport transfers, but that's small consolation. For many types of journey, flight-free transport simply can't compete on price alone. This is particularly true when there's an ocean to cross. In the late 2010s, it was easy enough to find a one-way flight from London to New York for £250 or less. Making a similar journey as a passenger on a cargo ship – Liverpool to Newark, say – typically costs around £1,200. The comparison is not strictly fair, of course. Unlike flying, the cargo ship voyage includes accommodation and meals for eleven days, with glorious views of the sea and sky thrown in.

Some argue that it is unacceptable for governments to keep the cost of flying artificially low through Open Skies agreements and tax breaks. Such initiatives encourage airlines to proliferate and leave them free to make decisions about routes, capacity and pricing that prioritize growth over the environment. In most countries aviation fuel, unlike road fuel, is not taxed and there's no VAT or sales tax on airfares. The UK charges Air Passenger Duty in an attempt to limit aviation emissions, but this isn't necessarily spent on green projects. Political and consumer campaigns are seeking to address these matters, but even if airfares end up rising, other fares won't automatically drop.

That said, flight-free travel can be great value if you stick to land transport and are a special-offer sleuth, making the most of advance booking discounts, seasonal deals, rail and bus passes and subsidized local transport. Luxembourg, for example, recently became the world's first country to make all standard-class bus, train and tram journeys free for both residents and visitors, in a push to get cars off the road and alert people to environmental issues. It uses public funds to make up the shortfall. Since early 2020, when this policy launched, other countries have been watching closely to see how well it works.

SAVE MONEY WITH TRAVEL PASSES

For lengthy multi-stop itineraries, train and bus passes offer huge savings, particularly if you score a discounted price through a travel agent. In Europe, Interrail and Eurail passes are deservedly popular, and some countries also offer affordable nation-specific passes – useful for in-depth trips. Other countries where transport passes are available include South Africa, Canada, USA, Australia and Japan. Always check the small print as validity options and exclusions vary. Sleeper services usually require reservations, for an additional fee. Some passes are only available to foreign visitors. For suggestions, see The Green Traveller's Directory, page 246.

Can flight-free travel get me where I want to go?

Assuming you have the time and funds to consider every available mode of public or independent transport, there's virtually nowhere on the planet that's totally out of reach of the flight-free traveller. But it would be wrong to pretend that travelling by land and water is always a breeze. Some major destinations present challenges.

Before planning a multi-country trip, check your government's travel advice concerning regions or land borders that may be unsafe. India, for example, can be difficult to reach overland: there are danger spots on the direct route from Europe via the Middle East, and only one relatively safe crossing point from Pakistan, at Wagah. You could take a circuitous route, looping through China, but it's important to bear in mind that the only way to travel through Tibet is on an organized tour, with a permit, and the border areas of Bangladesh and Myanmar can be dangerous.

Australia and New Zealand are more isolated than they look on the map: the only flight-free options are cruises – which would inflate your carbon footprint more than flights – or trips on private yachts or cargo ships. Australasian ports are fairly busy with freighters from Asia, Europe and the Americas, but have no ferries to or from Southeast Asia.

Similarly, Madagascar, Mauritius and Sri Lanka lie on busy freighter routes from South Africa and Singapore, but have no international ferries – there's not even a ferry between Sri Lanka and India. With no ferries and low cargo traffic, the Maldives, Seychelles, Pacific Islands and most of the Atlantic islands including Bermuda, Madeira, the Azores, Cape Verde and St Helena are awkward to reach, except by yacht. The Canary Islands are an exception, with ferries from southwest Spain.

While almost all Galápagos-bound tourists fly from mainland Ecuador, it's occasionally possible to bag a berth on a local cargo ship sailing from Guayaquil, a voyage of at least three days. Sadly, Easter Island is impossible to reach except by cruise ship or plane.

The Caribbean lends itself to island-hopping, but it's hard to join the dots by ferry. There are services from Florida to the Bahamas, but not beyond. The long-anticipated Florida–Cuba route is on hold indefinitely and there are no reliable ferries to Cuba from Mexico, or from Venezuela to the Lesser Antilles or Windward Islands. The best flight-free plan is to bring your own boat. Alternatively, arrive by cargo ship and join a flotilla holiday or charter a yacht, either bareboat or with a skipper and crew – a glorious way to enjoy the islands. Another option would be to aim for one of the few areas where ferries exist. They're very localized: you can hop around the British Virgin Islands, for example, or between Guadeloupe, Dominica, Martinique and St Lucia, or from Trinidad to Tobago.

The polar and subpolar regions including much of Alaska lie out of reach both of cargo ships and the Alaska Marine Highway ferries which ply the Gulf of Alaska. Much of Canada's northern hinterland is equally remote. Nunavut, for example, a territory covering more than 2 million square kilometres (770,000 square miles), has no roads or ferries leading to it; however there's a three or four month window in summer when relatively eco-friendly expedition cruise ships navigating the legendary Northwest Passage pass through. Further west, you can drive to the Yukon and the Northwest Territories from Alaska, British Columbia or Alberta – a superb road trip through pristine North American wilderness.

Antarctica, the Falklands, South Georgia, Greenland and Svalbard are supremely isolated – South Georgia doesn't even have an airstrip – but summer visits on expedition ships are possible. Iceland stands out as the most accessible subpolar island, with a weekly car ferry from Denmark, the Smyril Line MS *Norröna*. Sailing from Hirtshals via the Faroe Islands, it takes three nights to reach Seyðisfjörður on Iceland's beautiful, underrated east coast.

Accessible by ferry and road but far enough from Reykjavík to receive relatively few visitors, Iceland's Skaftafell National Park is superb for mountain treks, glacier hikes and ice climbing.

THE GREENEST WAYS TO TRAVEL

Hopefully it's not too long until low-carbon, eco-friendly transport becomes the norm. In the meantime, the best way to shrink your carbon footprint, minimize air and water pollution and boost your wellbeing is to choose your modes of transport with care.

CO_2e EMISSIONS PER PERSON PER KILOMETRE

Walking or wheelchair	0g		Petrol car (🚶🚶🚶🚶)	47g
Bicycle	0g		Motorcycle (🚶🚶)	50g
Electric motorcycle	0g		Local bus (hybrid)	79g
Electric/sailing boat	0g		Hybrid car (🚶)	107g
Electric car	0g		Motorcycle (🚶)	100g
Cargo ship	1g		Local bus (diesel)	120g
High-speed train	5g		Car ferry (🚶)	129g
Ferry (foot passenger)	18g		Plane flying 7,000km	146g
Hybrid car (🚶🚶🚶🚶)	27g		Plane flying 1,400km	153g
Long-distance bus	27g		Petrol car (🚶)	187g
Standard-speed train	37g		Plane flying 500km	244g
Car ferry (🚶🚶🚶🚶)	46g		Cruise ship	390g

NOTES

The figures quoted in this chart are grammes of carbon dioxide equivalent greenhouse gases (CO_2e) emitted per person per kilometre on an average journey.

These figures refer to direct emissions. For modes of transport that use fossil fuels and/or mains electricity that is not 100 per cent renewable, there are also indirect emissions associated with fuel production and supply. These may raise electric cars' total emissions well above zero.

In each case, the gases emitted may include carbon dioxide, methane, nitrous oxide, hydrofluorocarbons, perfluorocarbons, sulphur hexafluoride and nitrogen trifluoride. CO_2e is the universal unit of measurement to indicate the global warming potential of these emissions.

The figures make the following assumptions:

- Ships, trains, ferries, buses, planes: average occupancy.
- Cargo ships: each passenger's weight and energy consumption has a minimal impact on the ship's emissions.
- Car and passenger ferries: basic high-capacity ships carrying foot passengers, cars and freight, running on marine fuel. Emissions per passenger on ferries that don't carry freight tend to be higher.

- Cars/motorbikes: medium size; (🚶) indicates the number of people on board including the driver or rider. Emissions per person in large luxury vehicles can be 2.5 times higher than in small vehicles. Diesel cars produce 11 per cent less CO_2e emissions than petrol cars, but older models produce unhealthy levels of particulates.
- Long-distance bus: diesel-powered. At the time of writing, diesel is the norm, but hybrid and electric buses with lower emissions than high-speed trains are becoming more available.
- Planes: economy class. Emissions per passenger in business and first class are up to four times higher, reflecting the larger share of the cabin each seat occupies.
- Cruise ships: average capacity cruise ships or overnight ferries with cabins, running on marine fuel. On spacious ships with luxury facilities, the emissions per passenger are much higher.

All figures are approximate. Emissions for planes and land transport are based on UK government estimates (2020). Emissions for ships provided by Viking Line, Finland (2020).

Walking, cycling and sailing

Who says travelling from A to B need produce any emissions at all? For active types, there's a natural logic to giving up flying in favour of foot, pedal or wind power. If you're among them, you'll see flight-free travel as an opportunity, not a sacrifice.

Many nations have made itinerary-planning a doddle by creating waymarked walking and cycling trails. Leading along shorelines, rivers and canals, through forests or right across countries, they can take anything from a few hours to a few months to complete. Some provide back-to-basics camping experiences; others link places to stay.

Some of the most interesting trails, from a green perspective, are former railway lines. Typically, they're biodiversity corridors. Spain's Vías Verdes are great examples. The UK has dozens, such as the Phoenix Trail in the Chilterns and the gorgeously leafy Cuckoo Trail in East Sussex both of which lie close to functioning stations. In general, walking, cycling and trains are a good mix: the rules about bikes on trains vary, with some networks only allowing them outside peak hours, but it's easy to check online.

Sailing can be a great way to travel: skilled sailors can embark on long-distance voyages as crew by connecting with skippers via sailing clubs and websites. Holidays afloat are fun, too, whether you go it alone or join a guided trip with food, accommodation and baggage transport taken care of. The Caribbean, Greece, Croatia and Australia's Whitsunday Islands are all excellent for independent, skippered and flotilla yachting, and keen sailors can circumnavigate the UK in three months.

Electric cars, motorbikes and boats

Giving up flying doesn't mean giving up your freedom. With a zero-emissions electric vehicle or vessel, you can keep your timetable flexible and plan the perfect route, limited only by the availability of charging points – something that's rapidly improving.

Hybrid cars and boats are often a good compromise. If your only option is to travel by hybrid or, for that matter, something petrol-fuelled, remember that the more people on board, the lower each person's carbon footprint will be. With this in mind, it makes sense to consider hitchhiking or ride-sharing, perhaps using smartphone apps to help. But stay safe.

Cargo ships

When you have an ocean to cross, cargo ships are your friends. Their passenger quarters are reasonably comfortable and since your presence on board is incidental to the ship's main purpose, your journey's carbon footprint is effectively zero. However, berths are far scarcer than airline seats, and pricier. You'll need to book months in advance and be prepared for delays. Trickiest of all, for some, is that Wi-Fi is not a given during the long days at sea.

TURNING A NEGATIVE INTO A POSITIVE

Wiebe Wakker, who set out from his home in the Netherlands in 2016 to travel the world by electric car, coined the term 'range excitement' as an antidote to 'range anxiety'. Whenever his battery was running low, he simply threw himself into the fun of fixing up the next stopover.

Trains

When speed is of the essence, look no further than trains. Modern high-speed trains do exactly what their name suggests, while producing far lower emissions per kilometre or mile than regular trains, let alone planes. They also save you time and hassle by whisking you between city centres rather than suburban airports. In densely populated parts of Europe and Asia, they beat other modes of public transport in every respect, and are good value if you manage to bag passes or special offers.

If you've booked in advance, it's usually fine to arrive at the station as little as 45 minutes before departure. Once you're in motion, the distance melts away while you listen to music, read or snooze. Europe's fastest trains, Trenitalia's Frecciarossa 1000, travel at 360km/h (224mph). China's Shanghai Transrapid maglev train beats this, using strong electromagnets to lift and propel the train forward on a cushion of air at 431km/h (268mph). At the time of writing, a Chinese maglev train capable of 600km/h (373mph) has reached prototype stage.

Sleeper trains are highly practical, with numerous new services popping up, particularly in Europe, where overnight rail travel is booming. Splash out on a private compartment and you'll sleep far better than you would on an overnight bus. However, opting for a low-cost seat or bunk is a great way to make friends.

If there's a downside to trains, it's that organizing long-distance journeys can be a headache, particularly if you're travelling internationally, with multiple tickets and fare rules to juggle. Online booking services can assist, for a fee.

Tourist packages simplify things nicely. For romantics, the luxurious tourist trains such as the Venice Simplon-Orient-Express, Russia's Golden Eagle and South Africa's Blue Train are a must. And in a different realm to the maglev trains, technology-wise, are the gorgeous little vintage locomotives that puff their way along mountain tracks in Wales, Germany, Switzerland and elsewhere.

Long-distance buses

When the world's long-distance buses and coaches are all hybrid-powered or totally electric, they will be even greener than high-speed trains. It's a when, rather than an if, because they're already hurtling in this direction. In some parts of the world, such as South America and Australia, buses already beat trains for convenience, because their networks are far more extensive.

Buses are agile, often taking you very close to where you want to go, and modern models can be supremely comfortable, with good leg room and extras such as Wi-Fi and on-board refreshments. Some (particularly in Asia) are adapted for overnight travel, with reclining seats or even bunks. You can't expect to sleep soundly, but you will at least save the cost of a room for the night.

Car ferries, passenger ferries and expedition cruise ships

Ferries can be a lifeline to island communities, so are worth supporting: all too often services close because they can't break even. On long-distance ferries, your carbon emissions are lower than on cruise ships, because the total emissions are shared among a relatively large number of passengers, plus cars and freight.

Norway, a country with thousands of islands and fjords, is the world leader in adopting hybrid and all-electric eco-ferries. They're hugely successful, and many other countries are following suit: you'll already find them in San Francisco Bay, Brittany in France, Wellington in New Zealand, Denmark's Ærø Island, Scotland's Orkney Islands and England's Isle of Wight.

In destinations that tourists can only reach by ship, such as the Arctic and the Antarctic Peninsula, the greenest choices are small expedition cruise ships that have high ethical standards and eco-friendly protocols.

Ferries of the future

When teenage climate activist Greta Thunberg crossed the Atlantic by yacht in 2019, she made world headlines. Her zero-carbon sailing voyages, she shyly explained, were symbolic. She considered them essential to her campaign to tackle the climate crisis through global system change. 'I am not telling anyone what to do or what not to do', she said. 'I'm doing this to send a message that it is impossible to live sustainably today, and that needs to change.'

Thunberg freely admits that most transatlantic travellers aren't in a position to choose yachts over flights. For green-thinking individuals, not crossing oceans at all tends to be a more realistic goal. Nonetheless, in choosing not to fly but refusing to stop travelling, she sets an inspiring example.

What if, contrary to Thunberg's expectations, we wish to follow her lead? In the fast-moving field of slow travel, new options are emerging. Sail-hitching websites are popping up, allowing skippers and potential paying passengers to connect.

At the time of writing, Dutch company Fair Ferry is trialling sailing ship transfers around the Netherlands and beyond – even over long distances, to London, Copenhagen and New York. Meanwhile, ethical travel project VoyageVert is working towards creating a fleet of sailing ferries, capable of transporting hundreds of passengers across the globe. Perhaps a bright, wind-powered future lies ahead.

For travellers heading to the picturesque Lofoten archipelago in Norway, local ferries are a supremely relaxing way to arrive.

WHAT ABOUT CRUISE SHIPS?

Beware of cruise ships. Some cruise companies are bold enough to claim that they're eco-friendly or even carbon neutral, simply because they have streamlined the hulls of their ships, altered their fuel mix to reduce emissions, and offset the remainder by contributing to ocean-conservation programmes.

While improvements such as these are a step in the right direction, they don't make cruising green. As a mode of transport, cruise ships burn far more fossil fuel per passenger per kilometre (or mile) than aircraft, and as a place to stay, their carbon footprint is far higher than land-based hotels. Happily, there are signs that the industry wishes to continue making itself greener, though, by investing in ships powered by hydrogen, electricity or hybrid engines. Norway already has battery-powered day-cruise ships that are perfect for whale-watching in blissful silence.

THE BIGGER PICTURE

Making the switch to eco-friendly transport is the best way to make your trips greener. But why stop there? How can you make sure that your entire time away is just as responsible as your journey?

Stay somewhere sustainable

Choosing a green place to stay is getting easier all the time. Some hotels, B&Bs, homestays and campsites are eco-friendly from the foundations up, with certifications to prove it. Built or restored with sustainable, heritage-appropriate building materials, they're powered with renewable energy, furnished by local artisans and have water recycling systems and kitchen gardens. They're community-friendly, providing opportunities and avoiding any strain on local amenities or property values.

Some of the most rewarding places to stay keep their guests busy by organizing low-carbon activities such as yoga, crafting, wildlife-watching, walking, snorkelling or kayaking. Others offer learning experiences such as cookery, music or languages, allowing you to explore new horizons. They can also suggest good local restaurants, markets and shops, including vegetarian and vegan specialists.

For recommendations, hunt through guidebooks, search online for lists of eco-certified properties and contact travel companies that focus on responsible tourism.

Is your trip as green as it seems?

When it comes to sustainability, hotels, travel and tourism companies like to sing their own praises. Sometimes, their claims don't wash – in fact, they're greenwash. But how can you tell?

Find out whether they rely on carbon-offset schemes. While some schemes support worthwhile programmes such as the World Land Trust, investigations have revealed that many others don't actually result in a net reduction of emissions. For example, they plant trees that don't survive, or would have been planted anyway.

In any case, no offset scheme is a substitute for sound internal practices. Ask if you can see your holiday company's sustainability policy or ethical code, along with details specific to the journeys, accommodation or experiences that interest you most.

There are several questions you could ask. For example, does their operation benefit local people through profit-sharing, fair employment and welfare benefits? Does your trip or experience use eco-friendly transport, energy, food and materials to minimize its carbon footprint? Is it organic, water-wise and plastic-free? Finally, does it help fund local charities and contribute to the conservation of wildlife, habitats and traditional culture?

TREAT THE FAMILY TO AN ECO-FRIENDLY HOLIDAY

Camping holidays at activity resorts or green festivals can be very low-impact, with a stimulating variety of outdoor activities. Shambala in Northamptonshire, Green Gathering in Wales, We Love Green in Paris, Øyafestivalen in Oslo and Portugal's Boom in Alcafozes are all fun for kids. Another option is to look for farmstays or voluntourism programmes that interest you, perhaps at wildlife sanctuaries or heritage sites.

SEVEN WAYS TO TRAVEL BETTER

1. **Choose green transport**
   Allow time and money to minimize your journey's emissions, and try to do the same when planning activities – hiring a car that's electric rather than petrol powered, for example, or touring by bicycle instead.

2. **Consider the low season**
   Some destinations suffer from seasonal over-tourism; others struggle in the quieter months. Avoiding the busiest times helps relieve the pressure, and enables you to travel more spontaneously, scooping up discounts. Chances are, you'll receive a warmer welcome, too.

3. **Do your research**
   Research your destinations carefully, so that you arrive with some cultural and environmental understanding, ready to ask questions that will help you learn more. Be mindful of possible dangers, take health precautions, and make sure you're fully insured.

4. **Travel off the beaten track**
   In some developing countries, the tourist industry is highly localized, and young people drift out of their local workforce to seek jobs in the hotspots. To help restore some balance, try to visit far-flung community tourism enterprises.

5. **Be a model guest**
   As tourists, we're guests in someone else's home or neighbourhood. Politeness, consideration, humility, common sense, cultural respect and an open mind can go a long way towards making your stay a happy experience.

6. **Relax, but don't switch off**
   Travel light, and try not to be wasteful. When staying in hotels, B&Bs or homestays, find an opportunity to check that it's OK to re-use your towels and linen, recycle or compost your rubbish, steer clear of single-use plastic and avoid wasting food, water, heating or aircon, just like you do at home.

7. **Think local**
   Venture beyond the boundaries of your hotel, resort or tour group to try local food, buy from local traders, explore with local guides and visit local areas of natural beauty and biodiversity. By doing so, you help keep heritage, customs, conservation projects and small businesses alive.

GREEN CITY BREAKS

If a house is a machine for living in, then the best cities are buzzing, whirring factories of ideas. And right now, we need ideas aplenty. Creating well-connected, eco-friendly urban spaces to accommodate our growing populations is one of the greatest challenges we face. The cities that succeed – and are rich in history and culture, too – make superb places to visit.

Get to know Bristol by bike, boat and hot-air balloon

SOUTHWEST ENGLAND, UK

Far more than just a stopover on the way to the West Country and Wales, outdoorsy Bristol is worth a trip in its own right

Transforming British cities into cycling cities hasn't been easy. Their tightly woven, historic centres and car-centric post-war infrastructure present challenges aplenty. Bristol was the first to pull it off, back in 2008. As well as remodelling streets, the city invested in bike lanes, racks, training and a pioneering bike-share system.

It was the start of something big. Since then, Bristol's eco-credentials have earned it several green city awards. With abundant open spaces, low pollution and an active, eco-conscious population that embraces recycling, renewable energy, Fair Trade shopping and sustainable living, Bristol aims to become the UK's first carbon-neutral city by 2030.

For visitors, cycling is just one of many ways to enjoy the outdoors. Situated on the River Avon, 10km (6 miles) inland from the Bristol Channel, this maritime city has few remaining boatyards and warehouses, and has transformed its waterfront spaces into peaceful, car-free parks, ringed with footpaths and dotted with cafés, restaurants, pubs and places to hire bikes and watersports gear.

One of the docks in the historic Floating Harbour – an inner-city canal, deep enough to keep ships afloat at low tide – is home to Isambard Kingdom Brunel's magnificent *SS Great Britain*, now a museum. After your visit, stroll down to Cargo, a collection of ultra-hip mini restaurants, bars and shops in a stack of converted shipping containers. Pull up a stool, order a rainbow-coloured salad or a local craft beer, and you can watch the comings and goings on the water: yellow and blue passenger ferries pottering to and fro; rowers doing training laps; and kayakers, stand-up-paddle boarders, windsurfers and dinghies gliding along. There's usually a good range of age groups afloat, since Bristol funds free activities for kids and the over 50s.

Completed in 1864, Isambard Kingdom Brunel's Clifton Suspension Bridge was originally designed to carry horse-drawn traffic.

THE LOWDOWN

Best time of year: Any time. The Bristol International Balloon Fiesta takes place in August.

Plan your trip: Allow at least two or three days, perhaps as part of a tour of the West Country, the Cotswolds and Wales.

Getting there: Bristol is around 190km (120 miles) from London by road. The train journey from London Paddington takes around 1hr 40min. As the largest transport hub in southwest England, the city has direct rail and road connections from many parts of the UK.

WILD PLACE PROJECT

Calling itself a conservation project rather than a zoo, Wild Place, a 60-hectare (150-acre) park just north of Bristol, supports wildlife and habitat protection programmes in Africa and elsewhere. Its enclosures, linked by trails, are home to lemurs, cheetahs, giraffes and wolves, and there are free talks every day.

IN THE KNOW

South of the River Avon, Southville and Bedminster are Bristol's districts to watch. Come here for vegan junk food, farmers' markets, organic delis, microbreweries, independent bookshops and the Tobacco Factory, Bristol's best studio theatre.

Views of the river

To see more of the Avon, take a boat tour with Bristol Packet, whose vintage vessels motor along the Avon Gorge. Once you've sailed beneath Isambard Kingdom Brunel's elegant Clifton Suspension Bridge, the city recedes quickly, giving way to forested cliffs and leafy tidal wetlands. If you enjoy mountain biking, make a note to return to Ashton Court, west of the bridge, to charge along its woodland trails. And to admire the scenery from above, consider a hot air balloon trip, a Bristol speciality: the world's first eco-friendly solar-hybrid balloon was recently invented here.

Bristol's cultural scene is among Britain's best. With long-established universities, a multi-ethnic population and thriving creative industries, there's always something going on. The Welsh National Opera visits regularly, St George's Bristol hosts a superb range of concerts and the beautifully renovated Bristol Old Vic Theatre stages ground-breaking plays. There's a stimulating fringe scene, too, encompassing artists' open studios, gigs in pubs and walking tours with historical themes, from the whimsical to the macabre.

ALSO TRY

Brighton and Hove, East Sussex: Beautiful Regency architecture, a 6-km (4-mile) beach and access to the South Downs make this quirky, arty, green-voting city one of Britain's most attractive seaside destinations. Famously open to alternative lifestyles, it's excellent for vegetarian food, ethical shopping, live music and comedy.

King's Cross, London: Eurostar passengers can enjoy elegant shops and sustainable-cuisine restaurants in this revamped inner-city quarter, which is packed with urban planting and cutting-edge, energy-efficient features.

Edinburgh, Scotland: This historic city has embraced modern environmental principles. With green spaces, low pollution and invigorating slopes, it lends itself to strolling, even in winter. At the time of writing, Scotland currently has over 1,000 electric vehicle charge points that are free to use.

Brunel's Bristol-built SS Great Britain *was the world's largest passenger ship when it was launched. It crossed the Atlantic numerous times and made 32 voyages to Australia in its 1850s heyday.*

See Paris in a new light

NORTH FRANCE

THE LOWDOWN

Best time of year: Any time. Autumn and winter (October to February) are quietest.

Plan your trip: Allow at least a few days, either as a short break by bike, bus, electric car or train, or perhaps as part of a wider tour of France and western Europe.

Getting there: Paris is around 460km (285 miles) by road and sea from London via Kent, using the Eurotunnel motorail from Folkestone or a ferry from Dover. Alternative crossings with links to Paris include Portsmouth to Le Havre, Newhaven to Dieppe, Harwich to Hoek van Holland, Hull to Zeebrugge and Newcastle to Amsterdam. Direct Eurostar trains depart from London St Pancras (2hr 15min), Ebbsfleet (2hr 5min) and Ashford (1hr 50min). Train or bus journeys from elsewhere in the UK typically involve changing in central London.

Discover the French capital's green side by visiting out of season, and going where the locals go

Everyone knows that Paris is beautiful, elegant and cultured. Its galleries are stuffed with masterpieces, its boutiques are achingly stylish and its restaurants are divine. And many of us know that a landmark international agreement on climate action was signed and adopted here. But how does La Ville Lumière measure up as a responsible, eco-friendly, flight-free destination?

Helpfully, it lies at a crossroads. Take a look at Place Charles de Gaulle on a map: the sunburst of avenues radiating out in all directions from the Arc de Triomphe is a perfect metaphor for the city's excellent road and rail connections to western Europe, Eastern Europe and beyond (for flight-free travel ideas to Ukraine, Japan and Singapore, for example, see pages 194, 204 and 198).

As for eco-friendly initiatives: little by little, the home of the Paris Agreement is leading by example. France is closing coal-fired, oil-fired and nuclear power stations in favour of wind and solar energy and, in line with EU targets, Paris intends to be carbon neutral by 2050. It was the first European city to require supermarkets to donate waste food to charity, and it regularly stages climate-conscious events: under the Paris Respire initiative, for example, the first Sunday of each month is a car-free day in the city centre, allowing pedestrians, cyclists and scooters the run of the streets from 10am to 6pm.

Paris cherishes its green spaces, old and new. The Promenade Plantée, a 4.7-km (3-mile) disused railway line converted into a park, is delightful; created in the 1990s, it inspired New York's High Line and Chicago's Bloomingdale Trail. More recently, new riverside beaches and public parks have appeared, including the Parc Rives de Seine on the banks of the River Seine.

Buildings are getting greener, too, with living walls and rooftop gardens cropping up all over the city, particularly in Clichy-Batignolles, a future-facing eco-district in the 17th arrondissement, northwest of the centre. Energy-efficient architecture, cutting-edge water and waste systems and a pedestrian-friendly layout make this neighbourhood a wholesome place to live, while new-wave bistros, an arts centre and an organic market make it great to visit, too.

Hovering above the Parc André Citroën in the 15th arrondissement is the Ballon de Paris, a tethered sightseeing balloon, which doubles as an instrument to measure nitrogen dioxide, ozone levels and particles in the atmosphere. If the air quality alters, it changes colour: red is bad, yellow is mediocre and green is good.

IN THE KNOW

Keen to unlock 'le Paris des Parisiens'? The Balades Paris Durable website and Eco Walks smartphone app are full of useful clues, with ideas for self-guided walking tours in lesser-known neighbourhoods.

Paris in winter

If Paris has a downside, it's that its secrets are well and truly out. To explore it at its quietest and most serene, it's wise to avoid the main tourist season between March and September, skipping big events such as the Fête de la Musique in June, Bastille Day on 14 July and Fashion Week in late September. Opt for the crisp days of its autumn and winter instead, and rather than focusing on the big sights, visit lesser-known *arrondissements* and make your own discoveries.

This is a city where there's genuine joy to be had in the clichéd pursuit of living like a local. Settle into a self-catering apartment, download the Vélib' bikeshare app and enjoy simple pleasures: nip out for fresh bread from a neighbourhood *boulangerie*, drop into a church for a piano recital or just pedal along crowd-free streets, enjoying the architectural details. Temperatures from November to February generally hover below 5°C (41°F), but winter days tend to be drier than summer; if you wrap up well, there's no reason you can't linger over a coffee at a pavement table, and kick up the leaves in the local square.

To take the feeling of immersion a step further, you could book a homestay or hook into a social food network for a private dinner in a Parisian home. There's no better way to enjoy a taste of local *joie de vivre*.

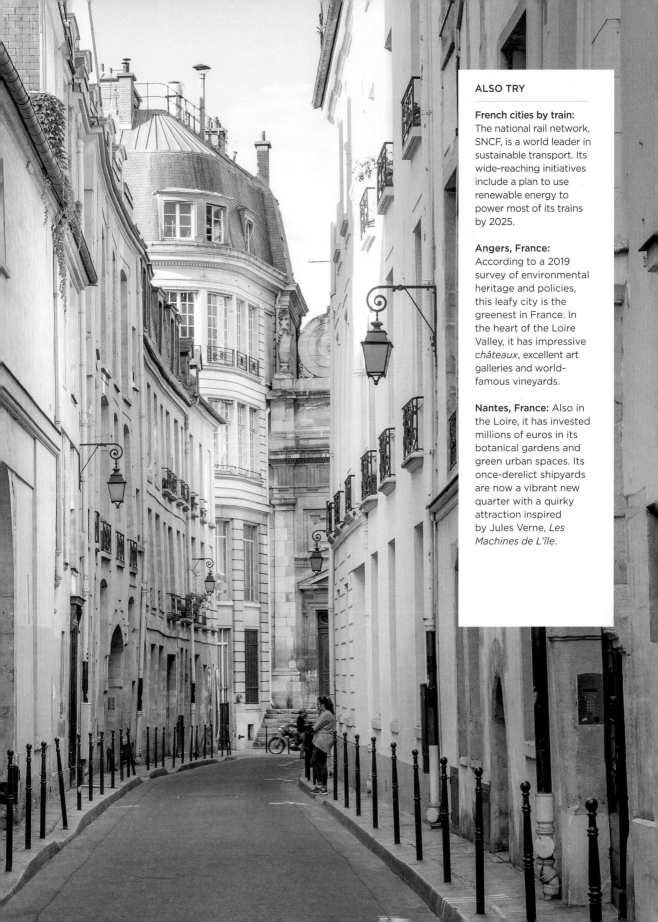

ALSO TRY

French cities by train: The national rail network, SNCF, is a world leader in sustainable transport. Its wide-reaching initiatives include a plan to use renewable energy to power most of its trains by 2025.

Angers, France: According to a 2019 survey of environmental heritage and policies, this leafy city is the greenest in France. In the heart of the Loire Valley, it has impressive *châteaux*, excellent art galleries and world-famous vineyards.

Nantes, France: Also in the Loire, it has invested millions of euros in its botanical gardens and green urban spaces. Its once-derelict shipyards are now a vibrant new quarter with a quirky attraction inspired by Jules Verne, *Les Machines de L'île*.

Decamp to Den Bosch, a hidden highlight of the Low Countries

NORTH BRABANT, THE NETHERLANDS

The birthplace of Hieronymus Bosch has beautiful nature reserves, a landmark cathedral and a unique artistic heritage

Best time of year: Any time. June to August is high season in the Netherlands.

Plan your trip: Allow at least two or three days, perhaps as part of a tour of the Low Countries by bike, bus, electric car or train.

Getting there: Den Bosch is around 480km (300 miles) by road and sea from London via France and Belgium. Rail routes from the UK typically begin with a leg to London St Pancras. From here, trains via Brussels, Rotterdam or Amsterdam take around 5hr 30min. Alternatively, travel to Harwich, Hull or Newcastle for a ferry, then continue by road or train.

Like that bargain bottle of wine that tastes like a vintage find, or that killer app that never crashes, Europe's hidden gems are elusive. There are plenty of reasons to avoid places that are vulnerable to mass tourism, seeking out culture-rich, queue-free destinations instead. But how do we go about finding them?

The Low Countries definitely deserve a second look. Beyond the hotspots of Amsterdam and Bruges, this region hides a wealth of secrets. The Dutch city of 's-Hertogenbosch, better known as Den Bosch, is a perfect example. Situated between Eindhoven and Utrecht, it's less than an hour by train from Amsterdam. One of the oldest cities in the Netherlands, it has genuine cultural clout, but is thoroughly underrated – if you haven't heard of it, you're not alone.

Den Bosch and its pleasant surroundings fit neatly into a flight-free tour of the UK and northern Europe. There are multiple ways to find your way to the Low Countries from the surrounding nations, including several routes from the British Isles. From England, for example, eminently viable options include zooming across by Eurostar, chilling out on a coach, loading your electric car onto the Eurotunnel shuttle or flinging your bike on a ferry.

As well as the well-known English Channel ferries from Dover to Calais in France, there are longer voyages that cross the North Sea in a day or a night, including the Stena Line service from Harwich in Essex to Hoek van Holland, P&O Ferries from Hull in Yorkshire to Europoort near Rotterdam and DFDS from Newcastle to IJmuiden near Amsterdam. Each of these Dutch ports is a manageable 115km (65 miles) or so from Den Bosch.

In line with EU climate goals, Den Bosch intends to be fully carbon neutral by 2050. It's already a cycling city, with excellent bike routes, and it has an active biodiversity programme with a special focus on butterflies.

IN THE KNOW

The bakers of Den Bosch have a wicked speciality that's sure to bowl you over: the *Bossche Bol*, a cream-filled chocolate profiterole, bigger than a tennis ball. It's perfect with coffee. Eat it with your fingers, and get messy.

Gables and Gothic flourishes

The first thing you'll notice as you stroll around is the city's architectural interest. The great artist Hieronymus Bosch, a lifelong Den Bosch resident born here around 1450, was doubtless inspired by the flamboyant style of Sint Jan's Cathedral, a Gothic masterpiece that was under construction throughout his lifetime. The Jheronimus Bosch Art Center in the city's medieval heart explores his work. The Noordbrabants Museum is stimulating, too, with regular exhibitions of contemporary art.

Out of town, there's much for walkers, cyclists, equestrians and wildlife-watchers to enjoy. Orchids and butterflies thrive in the lovingly restored wetlands of Vlijmens Ven, Moerputten and Bossche Broek, and the shifting dunes of the Nationaal Park De Loonse en Drunense Duinen (*pictured, above*) are less than half an hour away by bus. Nine times out of ten, you'll have plenty of space to yourself.

Handsome buildings, many with traditional stepped gables, line the pedestrian-friendly streets around De Markt, Den Bosch's spacious market square.

The Netherlands without the crowds: Besides Den Bosch, other underrated alternatives to Amsterdam include: The Hague, for art by Vermeer and Escher, waterside strolls and the beach. There are even more Vermeers in the pretty, walkable town of Delft, famous for its hand-painted pottery. Rotterdam, another stop on the Eurostar network, is great for European art, harbour bars and nightlife.

Belgium beyond Bruges: Instead of joining the hordes in Bruges, try Brussels, which receives less than half the volume of tourists. Fans of Art Nouveau, surrealism and graphic art love it, and it has more green space per resident than any other European capital. Or there's Ghent, with its extraordinary medieval buildings, green-thinking locals and veggie-friendly restaurants.

Cycle to your heart's content in Copenhagen

EAST DENMARK

Famously relaxed, with more bicycles than cars, the Danish capital is perfect for pedalling

With just over 1.3 million citizens, Copenhagen is large enough to be genuinely stimulating, but small enough to feel cosy – a favourite Danish attribute (remember all those books about *hygge*?). Liberal and progressive, it sets great store by the quality of life of its residents and visitors. Leading the race to become carbon neutral, it lies at the heart of a well-crafted network of pathways, bridges and ferries serving not just the city, but also its surroundings: low-lying islands, spacious beaches and flat, open countryside dotted with organic farms, cottages and castles. As such it's a natural choice for flight-free travellers, especially those who love cycling.

There are rental shops and bike-share stands all over Copenhagen, offering city, touring and electric bikes. The cargo bike, with a large container or child seat at the front, is a Copenhagen speciality: even the Danish royals have been spotted using them to ferry their kids to school.

There's no shortage of bike lanes and cycling highways, most of which are a pleasure to use. They include car-free bridges, such as the airy Cykelslangen (Bicycle Snake) across the Gasværkshavnen (Gasworks Harbour) waterway, and the elegant Inderhavnsbroen (Inner Harbour Bridge), right in the heart of the city. In the Nørrebro district, you can test your slalom skills in Superkilen, a paved public park designed to bring refugees and locals together, promoting ethnic tolerance. And when you're ready for a break from the urban bustle, you can pedal out to Beach Park, a 2-km-(1.2-mile)-long island long island with dunes and swimming areas, or Bellevue Beach, north of town.

THE LOWDOWN

Best time of year: April to August. High season is July and August.

Plan your trip: Allow a few days, perhaps as part of a tour of south Scandinavia by bike, bus, electric car, train or ferry.

Getting there: Copenhagen is around 1,250km (780 miles) by road and sea from London via France, Belgium and Germany. Rail routes from the UK typically begin with a leg to London St Pancras. From here, trains via Brussels or Amsterdam and Hamburg take around 15 hours or, with an overnight stop, just over 24 hours. Alternatively, travel to Harwich, Hull or Newcastle for a ferry to Belgium or the Netherlands, then continue by road or train.

It's not mandatory to wear a helmet when cycling in Denmark, so many Danes don't. However, they take cycling etiquette seriously enough for parents to feel confident in sending their kids to school and on educational trips by bike. If you hear a bell dinging, move to the right to allow the cyclist behind you to overtake. And never ride at night without lights, or take a left turn in one movement without raising an arm to indicate you're stopping at the corner of the road you're joining, as you could be fined.

Scrumptious smørrebrød and Nordic cuisine

After all that exercise, it's time to treat yourself. Yes, you'll find Danish pastries here, but under a different name – *Wienerbrød*, meaning Viennese. They're a Sunday breakfast treat. The classic Danish lunch is *smørrebrød* – an open sandwich of dark rye bread heaped with smoked salmon, marinated herring or salami and salad. Surprisingly, perhaps, the Danes export more bacon than they eat, with vegetables and fish far more popular.

At the time of writing, Copenhagen has eighteen Michelin-starred restaurants – almost as many as London, a city seven times the size. Fourteen of the city's restaurants have been awarded or commended under Michelin's inaugural Green Clover scheme for sustainability, and local chefs are refreshingly outspoken on environmental and ethical matters. Christian Puglisi of Relae, for example, who cooks with organic produce grown at his own experimental farm, feels the Michelin inspectors should make their sustainability criteria more stringent. And at the multi-award-winning Noma, René Redzepi, a chef who's passionate about foraging, believes that staff wellbeing is every bit as important as ethical sourcing. So tuck in, and if you can't clear your plate, ask for a doggy bag – a local campaign to fight food waste has made this perfectly acceptable.

Denmark has 11 National Cycle Routes, mostly safely separated from roads. Some pass gorgeous forests, farmland and coastal scenery. Route 9, from Helsingør to Gedser via Copenhagen, is a classic.

ALSO TRY

Oslo, Norway:
Like all the Nordic capitals, it's supremely sustainability conscious. Its new landmarks, the Munch Museum, Deichman Library and National Museum (due to open in 2021), showcase cutting-edge low-carbon architecture. The journey from Copenhagen takes around nine hours by train, or a little over 17 hours by ferry.

Stockholm, Sweden:
A stylish city that was the first European Green Capital, it still has a world-beating environmental record. Aiming to be climate-positive by 2040, it has an impressive 760km (470 miles) of bike lanes. By train, it's under five and a half hours from Copenhagen and under six and a half hours from Oslo.

Wander around Freiburg, a green city in the Black Forest

SOUTHWEST GERMANY

THE LOWDOWN

Best time of year:
March to June
and September to
November. High season
is July to August.

Plan your trip: Allow
at least two or three
days, perhaps as part
of a tour of Germany,
Switzerland and France
by bike, bus, electric
car or train.

Getting there: Freiburg
im Breisgau is around
870km (540 miles) by
road and sea from
London via Kent and
France. Rail routes from
the UK typically begin
with a leg to London
St Pancras. From here,
trains via Paris take
around 6hr 30min.
There's also a slightly
longer route via
Brussels, Cologne and
Frankfurt. Alternatively,
travel to Harwich, Hull
or Newcastle for a ferry
to Belgium or the
Netherlands, then
continue to Freiburg
by road or train.

Car-free medieval lanes and easy mountain hikes make Freiburg a glorious place to stretch your legs

With a university that was founded in 1457, Freiburg im Breisgau is an attractive city with strong opinions. In the 1970s, it became one of the birthplaces of the German environmental movement; today it's a modern eco-city and a stronghold for Germany's Green Party. According to meteorological experts, it's also officially the sunniest and warmest city in the country.

Freiburg's Old Town is an appealing jumble of cobbled medieval lanes, spread beneath the soaring spire of the Gothic cathedral, Freiburg Minster. Reconstructed after World War II, this quarter is still cooled by its original water supply, narrow open channels called *Bächle*. For panoramic views, hop on the Schlossbergbahn funicular for a three-minute ride to a hilltop viewpoint.

There are even wider panoramas from the Schauinslandbahn, Germany's longest cable car, which sweeps up the Black Forest mountain of Schauinsland from Horben, 11km (7 miles) south of town. For more time outdoors, the region has forest trails aplenty: it's around 10km (6 miles) from Freiburg Minster to the pilgrims' chapel of St Ottilien, or 14km (9 miles) between the villages of Himmelreich and Hinterzarten.

Green politics kicked off in Freiburg in the 1970s, with a successful campaign to block the construction of a nuclear power plant at Wyhl, northwest of the city. In 1980, the Green Party (Die Grünen) was founded in the nearby Baden-Württemberg city of Karlsruhe. Today, Freiburg aims to be climate-neutral by 2050 through ambitious construction, energy, transport, waste, nature and tourism initiatives, driven by the citizens themselves.

Locals have already stepped up to the plate by recycling fastidiously, making their homes more energy-efficient and buying direct from Baden-Württemberg's food producers.

PLANT-BASED CUISINE

In a region famous for *Wurst* and roast meat such as piglet, roe deer and *Sauerbraten* (marinated beef), vegans and vegetarians needn't feel alienated: Freiburg has dozens of restaurants that give veggies pride of place. Choices include Mediterranean, Mexican, Indian, Japanese and Afghan fare.

IN THE KNOW

Stay for two nights or more at a hotel or guesthouse belonging to the Schwarzwald Plus scheme, and you'll be given a card that allows free entry to more than 80 experiences in the Black Forest region, including castles, tours and shows.

In 1901, the Schwabentor (Swabian Gate), which dates back to the Middle Ages, was carefully adapted to allow Freiburg's first VAG electric trams to pass through. Today, the VAG network covers the entire city and runs on 100 per cent green energy.

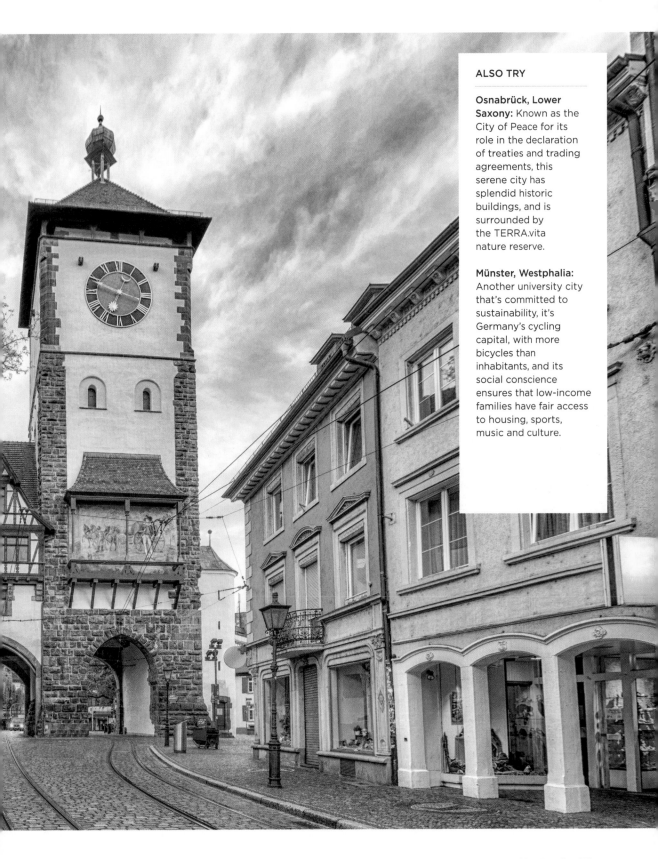

ALSO TRY

Osnabrück, Lower Saxony: Known as the City of Peace for its role in the declaration of treaties and trading agreements, this serene city has splendid historic buildings, and is surrounded by the TERRA.vita nature reserve.

Münster, Westphalia: Another university city that's committed to sustainability, it's Germany's cycling capital, with more bicycles than inhabitants, and its social conscience ensures that low-income families have fair access to housing, sports, music and culture.

Soak up the creative vibes in Dublin and Galway

REPUBLIC OF IRELAND

Cross the Emerald Isle from the Irish Sea to Galway Bay to discover a pair of cities that ooze good humour, colour and character

Best time of year:
June to September. High season is June to August and around St Patrick's Day, 23 March.

Plan your trip: Dublin and Galway are around 210km (130 miles) apart. The train journey from Dublin to Galway takes just over two hours; the bus just over three.

Getting there: Dublin is around 580km (360 miles) by road and sea from London via Wales. Travelling by train from London Euston and ferry from Holyhead takes around eight hours. At two hours, the Holyhead crossing is the fastest; there are also ferries from Liverpool (around 7hr 30min) and Douglas on the Isle of Man (3hr). Other options include the Cairnryan–Larne ferry from Scotland to Northern Ireland (2hr) or the Pembroke–Rosslare ferry from Wales to southeast Ireland (4hr), continuing by road or rail.

Island countries can be tricky to get to without flying, but that's no reason to omit the island of Ireland from your wish list. It has more sea connections than you might realize, with ferries from nine ports in Wales, England, Scotland, the Isle of Man, France and Spain. Dublin, Ireland's likeable capital, is just a short hop across the Irish Sea from north Wales: if you caught a train out of London around noon, you could be sipping a Guinness near Dublin's Ha'penny Bridge by 8.30pm and lunching in Galway the next day.

The Irish Sea is notoriously choppy, but modern ferries such as the *Dublin Swift* from Holyhead do their best to make the crossing smooth and speedy. The *Swift's* well-designed interior is far more spacious and comfortable than a plane's.

Both cities are members of the UNESCO Creative Cities Network. Dublin, birthplace of giants such as James Joyce, Samuel Beckett and Oscar Wilde, is listed as a City of Literature, while Galway is listed as a City of Film with a thriving movie, television and animation industry. Each has much to offer: Galway has two of Ireland's best cinemas, and Dublin's libraries and literary museums are superb. Both cities enrich their cultural climate with regular festivals and events.

Dublin, a well-read capital with high-tech goals

With handsome Georgian architecture, a famous university and a burgeoning tech industry featuring such giants as Google, Amazon, Facebook and Twitter, Dublin is a city with a spring in its step. It's walkable and cycle-friendly, with plenty of green spaces including St Stephen's Green, Herbert Park and the beautifully landscaped War Memorial Gardens, designed by Sir Edwin Lutyens.

Home to the inspiring National Library of Ireland – the most comprehensive collection of Irish fiction, non-fiction and documentation in the world – the city airs its latest literary tastes during its annual International Literature Festival, held in May. Thanks to the thriving Made In Ireland movement and the locals' taste for all things vintage, artisan and one-of-a-kind, this is also a fun place to shop. There's a host of independent boutiques and a lively flea market, held on the last Sunday of each month from April to September at The Digital Hub on Thomas Street.

Spanning the River Liffey, the much-loved Ha'penny Bridge sometimes serves as a backdrop for events in the International Literature Festival Dublin, held in May.

GALWAY INTERNATIONAL ARTS FESTIVAL

Held in July, Galway's two-week festival is one of Ireland's largest and liveliest. It features big-name rock, world and contemporary musicians, street spectacles, theatre, art installations and talks.

On Galway's Long Walk, a terrace of quayside houses looks southwest towards the former fishing village of The Claddagh.

Galway, a cool city where movies matter

The journey from Dublin to Galway leads across the rolling pastures of Leinster and Connacht. If you're lucky with the weather, you'll see why film-makers find the west-coast light so inspiring. To binge on movies, head for Galway's Eye Cinema or the strikingly modern Pálás.

The city's youthful energy and sophisticated aesthetic really come to the fore in its contemporary restaurant scene. At Aniar, chef J P McMahon whips up seasonal specialities such as brill with sea beet, a foraged seashore plant with succulent green leaves. Enda McAvoy's Loam, the winner of Michelin's inaugural Sustainability Award for Great Britain and Ireland, serves original dishes such as scallops with cauliflower and bottarga (cured roe). Working in an open kitchen, the team uses every possible part of each plant and animal, composting the rest. To discover more local treats, from oysters to strawberry tarts, book a food-themed walking tour.

Irrepressibly vivacious, Galway is also an excellent place to sample one of Ireland's favourite pastimes – music-making in pubs. You can't beat 'the *craic*' at the traditional bars in the Latin Quarter and West End. Monroe's Tavern, Taaffes Bar and Tig Cóilí, for example, host live bands every night. Come prepared to join in.

ALSO TRY

Glasgow, Scotland:
A UNESCO City of Music, it hosts arena concerts and musicals at the SSE Hydro or the landmark SEC Armadillo, designed by Foster + Partners in Glasgow's Scottish Event Campus (SEC). For more intimate recitals, visit the Mackintosh Church in Queen's Cross.

York, England:
The world's first UNESCO City of Media Arts has revived its ancient tradition of Mystery Plays, started a modern tradition of son et lumière and created a new festival, York Mediale, featuring music, speech and radical art.

Buskers perform in the pedestrian streets of Galway's medieval Latin Quarter.

Flit between Berlin and Vienna by rail

GERMANY AND AUSTRIA

Linked by sleeper train, the German and Austrian capitals are perfect for night owls and culture vultures

Interrailers have long been wise to the virtues of Europe's night trains. Even if you have a rail pass that limits you to just four, five or seven travel days in one month, sleeper services can save you both time and money. An overnight journey requires a small reservation fee, but uses up only one of your travel days: the day the train departs, not the day it arrives. For this, you can zoom comfortably between cities, saving the cost of a room for the night.

Berlin and Vienna are connected by the excellent Nightjet network operated by ÖBB (Österreichische Bundesbahnen, Austrian Federal Railways). You'll leave Berlin in the early evening and will arrive, with no change of time zone, in time for breakfast in a Viennese café: coffee, *Semmeln* (bread rolls) and apricot jam if you're in the mood for something traditional, or eggs, avocado, smoked salmon and oysters if you're feeling lavish. According to ÖBB, a return trip saves you around 183kg (0.2t) of greenhouse gases compared to taking the plane.

Germany's diamond in the rough

Berliners are mad-keen cyclists. Competing bike-share schemes come and go, and some are better than others, so it's best to ask a local which to use. There are mopeds for hire, too. It's fun to soak up the city centre's rich history on two wheels, hopping between unmissable spots such as the Brandenburg Gate, Reichstag and Hackescher Markt, with its cafés and edgy boutiques.

Flea markets and food trucks make Berlin a sustainable shopper's goldmine. Two of the best locations for pre-loved fashion, art, vinyl and street eats are Boxhagener Platz and the Mauerpark, where stalls set up every Sunday. For vintage shops, speciality coffee, global cuisine and hipster nightspots, don't miss gritty Kreuzberg, one of the city's most multicultural districts.

Completed in the 1960s, the landmark Fernsehturm (TV Tower) has come to symbolize a reunited Berlin. On a clear day, the view from its observation deck stretches for 42km (26 miles).

THE LOWDOWN

Best time of year: Any time, particularly May to October. Winter brings daytime temperatures below 5°C (41°F). High seasons are Christmas, New Year and June to August.

Plan your trip: Berlin and Vienna are around 640km (400 miles) apart, taking around 8hr 30min by high-speed train or 12hr 40min by night train.

Getting there: Berlin and Vienna are around 1,100km (710 miles) and 1,470km (910 miles) respectively by land and sea from London. Eurostar trains from London St Pancras to Brussels or Amsterdam connect with trains to Berlin (high-speed) and Vienna (high-speed via Frankfurt, or overnight direct). The journey from London to Berlin takes around 10hr 20min. To Vienna, it's around 13hr 40min (or 17hr 30min by Eurostar and sleeper).

Nightjet trains have
four classes: Sitzwagen
compartments for six
people sitting upright;
Liegewagen couchettes
for four or six in bunks;
and Schlafwagen sleeper
cabins for one, two or
three, standard (with a
basin) or deluxe (en-suite).
When choosing, bear
in mind that you're
unlikely to sleep well in
a Sitzwagen, since you
can't lock away your
belongings or lie down.
Liegewagen and
Schlafwagen passengers
receive bedding. If you're
sharing a couchette
or cabin with strangers,
the etiquette is to sleep
in your clothes.

Austria's polished cultural capital

Vienna may not be Europe's geographical centre (pub quiz wizards claim that's in Lithuania), but thanks to the scope and efficiency of the Austrian rail network, it's arguably the continent's pivot point. Wien Hauptbahnhof (often abbreviated to Hbf), the city's sleek central station, is a major international and mainline rail hub: soon after it opened in 2015, it was handling 145,000 people per day. Only Moscow comes close in having connections to so many other European capitals.

Culturally, Vienna is highly influential, famous for its associations with Mozart, Beethoven, Egon Schiele and Gustav Klimt. Masterpieces by Schiele and Klimt are displayed in several Viennese art museums. For all its baroque splendour, Vienna has a contemporary gloss, too, with impressive modern architecture by Dominique Perrault and Coop Himmelb(l)au and hipster hangouts galore in Neubau, the bohemian 7th District.

IN THE KNOW

European sleeper trains
are often extremely long,
so look for a carriage
plan on the platform.
Sector signs hanging
overhead will help you
wait in the right spot
and calmly board by
the correct door.

Vienna's Hundertwasserhaus was designed by artist Friedensreich Hundertwasser and architect Josef Krawina in the 1980s.

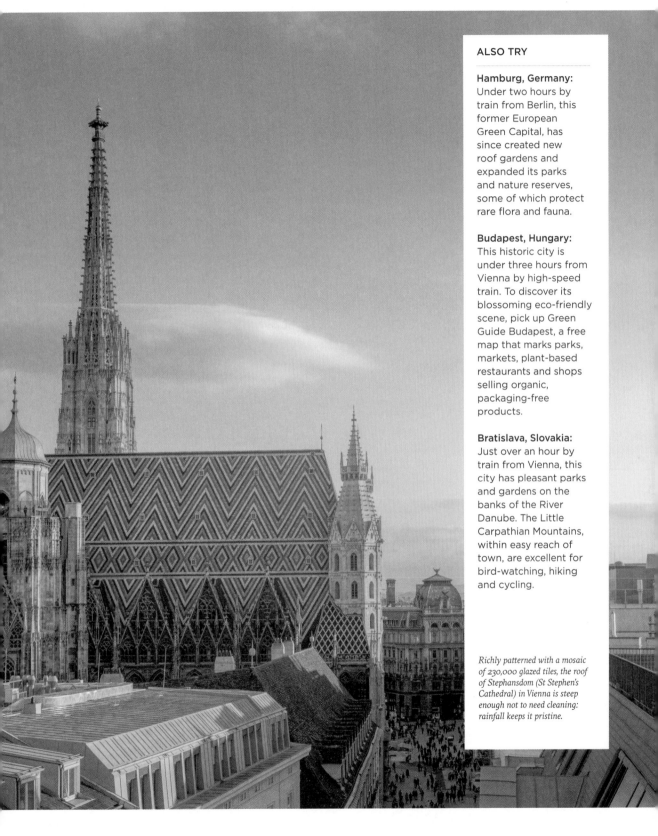

ALSO TRY

Hamburg, Germany:
Under two hours by
train from Berlin, this
former European
Green Capital, has
since created new
roof gardens and
expanded its parks
and nature reserves,
some of which protect
rare flora and fauna.

Budapest, Hungary:
This historic city is
under three hours from
Vienna by high-speed
train. To discover its
blossoming eco-friendly
scene, pick up Green
Guide Budapest, a free
map that marks parks,
markets, plant-based
restaurants and shops
selling organic,
packaging-free
products.

Bratislava, Slovakia:
Just over an hour by
train from Vienna, this
city has pleasant parks
and gardens on the
banks of the River
Danube. The Little
Carpathian Mountains,
within easy reach of
town, are excellent for
bird-watching, hiking
and cycling.

*Richly patterned with a mosaic
of 230,000 glazed tiles, the roof
of Stephansdom (St Stephen's
Cathedral) in Vienna is steep
enough not to need cleaning:
rainfall keeps it pristine.*

Green City Breaks 61

Fall in love with the soul of Iberia in Madrid and Lisbon

SPAIN AND PORTUGAL

Cruise between Spain and Portugal's enjoyable, arty capitals on a relaxing train journey

Atocha, Madrid's primary railway station, takes green travel to unparalleled heights. Instead of a regulation concourse dotted with coffee kiosks, it has two zones: a modern terminal with access to the platforms, and a 19th-century hall filled with tropical plants that would put many a botanical garden to shame. Glass skylights between the wrought-iron roof arches flood the space with light, and cafés invite you to relax as you wait for your train.

Sadly the high-speed rail link between Atocha and Lisbon, proposed in the 1990s, never materialized, and at the time of writing, the daily Trenhotel Lusitania sleeper train between the Iberian capitals has been withdrawn. It's far faster to travel across La Mancha, Extremadura and Alentejo by car or bus than by the remaining train services, which travel by day and require a couple of changes. However, if you value comfort over speed, you'll love this rather lengthy intercity railway journey. Cruising across the open landscapes of Spain via Badajoz and Entroncamento in air-conditioned trains, it's a leisurely trip, which gives you ample time and space to stretch out and read, chat or snooze.

Trains from Madrid, Badajoz and Entroncamento roll into Lisbon's 1990s Oriente station, northeast of the city centre. Exuding a different kind of glamour to Atocha, it was designed by the award-winning Spanish architect Santiago Calatrava and its roof is a forest of beautiful geometric arches. The Oceanário de Lisboa, a superb aquarium with no captive dolphins or whales, is nearby. If you prefer, you could stay on board for another eight minutes and alight at Santa Apolónia, Lisbon's 19th-century central station.

THE LOWDOWN

Best time of year: Spring (March to May) and autumn (September to November) bring the most comfortable weather.

Plan your trip: Madrid and Lisbon are around 625km (390 miles) apart, taking 11 hours by regional trains.

Getting there: Madrid is around 1,720km (1,070 miles) by road and sea from London via Kent and France. Eurostar trains from London St Pancras to Paris connect with high-speed trains via Barcelona to Madrid, taking around 14hr 10min. Alternative routes from the UK include ferries from Plymouth or Portsmouth across the Bay of Biscay (18–24hr), docking at Santander or Bilbao, which are under five hours north of Madrid by regional train. For direct routes between the UK and Lisbon, see page 92.

Soaring over Lisbon's lively and characterful Alfama district is the white dome of the Panteão Nacional (National Pantheon), which houses the tombs of Portuguese notables, from presidents to stars of football and fado music.

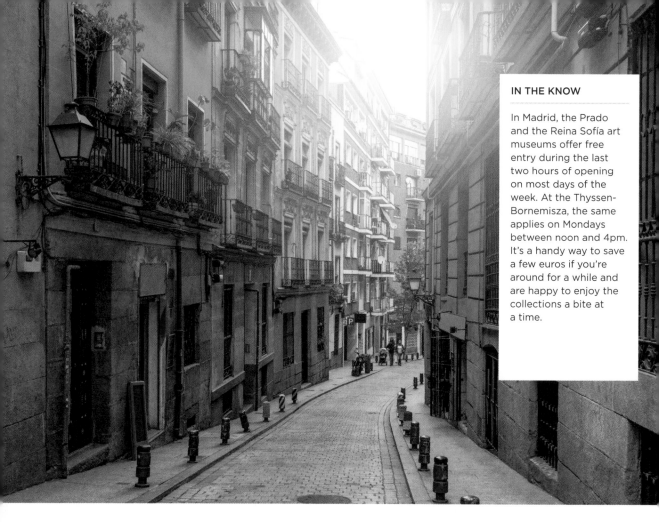

EAT LIKE A LOCAL IN LISBON

Lisboetas (or Alfacinhas, as Lisbon's locals are sometimes known) will tell you that, when picking a restaurant, high prices don't necessarily guarantee good food. It's better to choose by reputation, so quiz your host, or try hunting through local blogs for recommendations. Most importantly, don't expect dinner before 8pm: anything earlier is just a snack.

Dine out on art in Madrid

For art lovers, Madrid is a feast. The golden triangle formed by its three great art museums, the Prado, Reina Sofía and the Thyssen-Bornemisza, is a treasure trove of works by Picasso, Dalí, Goya, Velázquez, Caravaggio, Rembrandt and many others. There's more than enough to keep your eyes out on stalks for several days.

Despite its many attractions, Madrid hasn't suffered as badly from mass tourism as Barcelona, partly because the capital's planning department has placed strict limits on casual holiday lets. The city is particularly pleasant in autumn, when the heat has ebbed away and the plane trees in Retiro Park burst into colour.

Bar-hop by tram in Lisbon

The supremely liveable city of Lisbon was European Green Capital in 2020. Driving is discouraged here, unless in an electric vehicle. The city has one of the world's largest networks of street-side charging points, but why would you drive, when you can take a vintage tram instead, or jump on an electric bike at a bike-share stand?

Bairro Alto is the neighbourhood that's most loved by locals and visitors alike. It's quiet by day and lively by night. With cobblestones, street art, welcoming bars, great restaurants and live music, there's always something going on. It's also the best place to buy *pastéis de nata* (crunchy custard tarts) outside their mother lode, nearby Belém.

ALSO TRY

Sud Expresso: Named after the classic sleeper train that until recently ran diagonally across the peninsula, this route wends its way between Lisbon, Madrid and the Basque towns of San Sebastián in northeast Spain and Hendaye on the French-Spanish border. Hendaye is under five hours by direct train from Paris and a slow but scenic train ride along the Spanish coast from the ferry ports of Bilbao and Santander, making it a useful stop in a journey between Portugal and England or Ireland.

El Transcantábrico Gran Lujo: Spain's original luxury hotel-train travels along the Biscay Coast from San Sebastián to Santiago de Compostela. Bilbao, with its landmark Guggenheim Museum, merits a visit in its own right. Other highlights of the eight-day trip include Santillana del Mar and the Picos de Europa.

Al Andalus: This luxury hotel-train explores the heritage highlights of Andalucía, southern Spain, spending a week touring Seville, Jerez, Ronda, Granada and Córdoba. For more time in each city, you could drive a similar route in a hired electric car.

Stimulate your senses in Tangier, Casablanca and Marrakech

MOROCCO

Hop on Morocco's new bullet train and absorb the energy of the country's hectic urban centres

Wave goodbye to England, France, Spain or Portugal today, and you could be in Morocco tomorrow. The cultural connections between western Europe and North Africa, which date back centuries, remain as strong as ever, partly because it's so easy to leap between these two very different worlds. Ferries across the Strait of Gibraltar are fast and frequent, making Morocco's vibrant collection of cities an excellent target for a flight-free adventure, especially now that its high-speed railway is off the starting blocks.

Al Boraq, launched in 2018, is Africa's first bullet train. With a top speed of 320km/h (200mph), it has reduced the journey between Tangier and Casablanca from around five hours to just over two, and it will travel even faster once track improvements are complete. With smart double-decker carriages, gleaming stations and bargain fares, the high cost of creating this new railway has, inevitably, attracted controversy, but for visitors, it's a boon. The plan is to extend it to Marrakech, Agadir and other cities including Essaouira. In the meantime, if you're bound for Marrakech, you can change at Casablanca for the onward journey, or travel all the way down from the Strait of Gibraltar on the night train from Tangier.

Meet me in the medina

An overland journey through Spain – admiring the Mudéjar architecture in Zaragoza, Córdoba and Granada, for example – sets you up nicely for your arrival in Tangier. Romanticized in dozens of novels, poems, paintings and films, the city is soaked in North African cool. As your ferry motors into port, you're greeted by a jumble of whitewashed buildings that echo the Moorish towers and arches found all over Spain, but are unmistakably Moroccan. Like many coastal cities, it has a raffish side – there are touts, hustlers and extensive dilapidation to deal with – but if you can cope, the rewards are great: watching bakers and weavers at work, sipping mint tea or just feasting on classic Moroccan, French and Spanish food.

Once a Phoenecian colony, Tangier has served as a gateway to northwest Africa for over 2,000 years.

THE LOWDOWN

Best time of year: March to May and September to October. Summer and winter can bring extreme temperatures.

Plan your trip: Allow a minimum of ten days for a return trip with at least two nights in each city. It's around 340km (210 miles) from Tangier to Casablanca, and 240km (150 miles) from Casablanca to Marrakech. Bullet trains from Tangier to Casablanca take 2hr 10min; regular trains from Casablanca to Marrakech take 2hr 40min.

Getting there: Tangier is around 2,500km (1,500 miles) by road and sea from London via Kent, France and Spain. By train, travel from London St Pancras via Paris and Barcelona to Algeciras in southern Spain (around 20h 40min), then transfer by bus to Tarifa (30min) for the ferry across the Strait of Gibraltar (1hr). The journey can be done in two days, with an overnight stop. Buses from London via Paris take around 42 hours including ferries.

EXPLORE WITH
A GUIDE

If you're daunted by touts, traffic and sensory overload, hire professional guides for walking tours of the Moroccan cities you visit. As well as keeping hustlers at arm's length, their local knowledge and interpretative skills will illuminate what can otherwise be a confusing and even intimidating experience.

South to Casablanca and Marrakech

Despite its silver-screen fame, Casablanca is often overlooked by tourists. However, that may change now that the new bullet train links it to Rabat and Tangier. While it lacks the rich colours and atmosphere of other Moroccan cities, it has fabulous Art Deco buildings, a spectacular 1990s mosque and bags of contemporary character. Yes, it has Atlantic beaches, but it plays down its seaside city image: it's very much a modern metropolis with a business head on its shoulders.

Marrakech, in all its restless, pungent, cacophonous glory, is an invigorating place to spend a few days. Like every Moroccan city, it has a buzzing medina, and its famous square, Djemaa el-Fna (*pictured, right*), is a living tapestry of fortune tellers, fruit sellers, street-food stalls and tourists. But there are quiet corners, too. The famous Majorelle Garden with its vivid blue cubist features is serene in the early morning, before the Instagrammers arrive; and the city's many renovated riads have beautifully tranquil courtyards.

IN THE KNOW

When shopping in Moroccan souks, haggling over prices is standard practice. If it's clear you're a tourist, the opening price may be ten times the sum the vendor is willing to accept. Forget any concerns that you, as a relatively wealthy foreigner, should simply pay or go. This is commerce, not charity, so engage, and enjoy the game.

ALSO TRY

Rabat: The Moroccan capital, another stop on the Al Boraq railway, is smaller than Casablanca, and considerably calmer: even the medina is less chaotic than most.

Fes: This underrated Moroccan city has a richly historic, spiritual and cultural atmosphere. It has a wealth of beautiful buildings, such as the superbly ornate University of al-Qarawiyyin.

Essaouira: An independent traveller's favourite, this breezy city on Morocco's Atlantic coast has a lively watersports scene and fascinating musical traditions (see page 112).

Dose up on architecture in Helsinki, Tallinn and St Petersburg

FINLAND, ESTONIA AND RUSSIA

THE LOWDOWN

Best time of year: May to September. However, if your heart is set on cross-country skiing and the northern lights, December to March is best. High season is July and August.

Plan your trip: Frequent ferries connect Helsinki to Tallinn (2hr), and high-speed trains run from Helsinki to St Petersburg (3hr). The overnight ferry from St Petersburg to Helsinki and Tallinn takes 13-14hr.

Getting there: The most relaxing route from the UK to the Gulf of Finland is to travel to Stockholm (24–36hr from London; see page 108) for a ferry across the Baltic to Helsinki or Tallinn (18hr). Tallinn and St Petersburg are around 2,610km (1,620 miles) and 2,800km (1,750 miles) respectively from London by road and sea via Poland. Train-hopping from St Petersburg to London takes around 41hr. There are overnight buses from Tallinn to Warsaw, with rail and road connections.

The Gulf of Finland's fascinating trio of cities lends itself to a circular trip, travelling by train and ferry

In the long, light days of summer, when the Gulf of Finland is smooth as a lake, its shore is a serene place to be. Three cities face these waters: the Finnish and Estonian capitals of Helsinki and Tallinn, and St Petersburg, Russia's cultural centrepiece.

The Gulf of Finland is the eastern arm of the Baltic Sea, a vast inlet of the Atlantic that Scandinavians, Germans and Eastern Europeans fought over for centuries. Winter hits this region hard, but the Gulf helps keep temperatures moderate. St Petersburg's shores are scattered with ice between December and April, and in harsh seasons the entire gulf can freeze, but these days it's rarely solid enough for shipping to be cancelled.

Ferries and cargo ships are constantly plying these waters, and in recent decades, maritime pollution, agricultural run-off and urban waste have taken their toll on the marine environment. But thanks to joint initiatives from Finland, Russia and Estonia, it's now recovering. Grey seals are increasing in number, and Baltic ringed seals are clinging on, despite suffering habitat loss due to climate change – they breed on sea ice.

Helsinki, an architectural showpiece

Set on the gulf's northern shore, Helsinki is Europe's busiest passenger ferry port. The most popular route, the two-hour hop south to Tallinn, accounts for over eight million trips each year. If you're staying in Tallinn, you could easily visit Helsinki for a day, or vice versa, but both cities definitely merit a longer stay.

Charming Art Nouveau details, Neo-classical terraces, the neo-Gothic Design Museum and airy modern buildings by Alvar Aalto and other notable architects give Helsinki a delicate character. The city has some truly original structures, such as the Amos Rex (*pictured, left*), an avant-garde art museum that opened in 2018, the Kiasma Museum of Contemporary Art and the origami-like Löyly, a public sauna built of timber, floating over the waterfront. The city centre is walkable, and visitors are encouraged to explore lesser-known neighbourhoods, enjoying glassware galleries, futuristic workshops and top-ranking restaurants serving low-waste, hyper-local, plant-based cuisine.

If you love forests and cross-country skiing, visit in winter and allow some time up-country; if coastal scenery and kayaking appeal more, come in summer and take the ferry west to Turku near Finland's southwestern tip, or to nearby Mariehamn and Långnäs in the Åland Islands, a beautiful archipelago of 7,000 or so islets between southwest Finland and Stockholm.

CARBON-NEUTRAL HELSINKI

Helsinki is a strong contender in the race to become Europe's first carbon-neutral capital. It's investing in cycling infrastructure, zero-emissions buses and electric-car charging points. Imaginatively, it also believes that not everything needs to be owned: the landmark new Oodi library offers 3D-printing, sewing machines, a photo studio, tools and games to borrow as well as books.

IN THE KNOW

If you arrive in St Petersburg by ferry on a trip booked through an authorized travel agency, you're permitted to stay for 72 hours without a Russian visa.

St Petersburg's Church of the Saviour on Spilled Blood, now a museum, contains over 7,500 square metres (9,000 square yards) of mosaics.

In Tallinn's old town, the tucked-away lane of Katariina Käik is home to Katariina Gild, a craft collective where visitors can watch artisan glass-blowers, jewellers, weavers and potters at work and buy from them direct.

Medieval Tallinn and elegant St Petersburg

Beautifully preserved, Tallinn's quaint Old Town is a UNESCO World Heritage Site, its skyline bristling with the spires of opulent churches. Its knot of medieval streets dates back to the 13th century, when this was an outpost of the Hanseatic League, a trading federation with centres in Lübeck and Gotland. Tallinn's Medieval Days festival, held in July, celebrates the past with costumed musicians and epic feasts.

For art lovers, Kumu Art Museum is a highlight: alongside a permanent collection of Estonian art, it hosts stimulating exhibitions of contemporary works by international painters and sculptors.

Head east from Helsinki by train, or hop on Moby SPL's overnight ferry, the Princess Anastasia, which links the Gulf of Finland cities twice a week, and you'll reach the glittering city of St Petersburg. Founded in 1703, it's a stripling compared to Helsinki and Tallinn. Stuffed with grand 18th-century palaces designed to broadcast the wealth and might of imperial Russia, its gilded extravagance is enough to leave you breathless. The crowning glory, the monumental Winter Palace, is home to the Hermitage Museum, which flaunts a wealth of treasures based on the private art collection of Empress Catherine the Great.

ALSO TRY

Riga, Latvia: Another UNESCO World Heritage listed Baltic city, the Latvian capital has splendid Gothic and Art Nouveau architecture.

Moscow, Russia: Travel from Helsinki to Moscow on the revamped Russian Railways Lev Tolstoy, a night train with a restaurant car serving regional cuisine. Come morning, you'll pull into the 19th- century Leningradsky Station, ready for a deeper immersion in Russian culture.

BLUE
SEA
THINKING

Is your ideal beach rocky or sandy? Do you prefer the skies blue or star-spangled, and the water stormy or serene? Whatever floats your boat, there's much to enjoy where the land meets the sea. Best of all, there's no real need to jet off to the sun, even if you're hankering for a taste of island life. There are pretty coves and flower-speckled clifftops a train trip or ferry ride away. You just need to know where to look.

Tour Scotland's western isles, from Arran to Skye

WEST SCOTLAND, UK

Whether you're a whisky connoisseur or you simply like chasing rainbows, you'll find island-hopping in the Hebrides a delight

When it rains on North Ayrshire and the Inner Hebrides, it doesn't hold back. But when it shines, it sparkles. The air is so clear in these parts that distant headlands appear freshly polished and rainbows seem ludicrously vivid.

Make no mistake: when travelling up the coast from the Firth of Clyde by ferry, you're likely to encounter every type of chilly, wet weather the season can throw at you. Chat with the locals, and you'll soon learn the words *dreich* (wet and dull), *haar* (sea mist), *snell* (biting wind) and *drookit* (a total drenching). But the moment the sun appears and those technicolour *watergaws* (patches of rainbow) blaze overhead, it'll all feel totally worth it.

Set out on the whisky trail

The Isle of Arran and the Inner Hebrides islands of Islay, Jura, Colonsay, Mull, Iona and Skye link together as naturally as beads on a string. Begin your 583-km (362-mile) journey in the North Ayrshire town of Ardrossan. Arran, 55 minutes away by ferry, has the grand Victorian estate of Brodick to explore, complete with castellated mansion and spacious grounds, home to rhododendrons and photogenic red squirrels. To summon the spirits of ancient times, visit the 4,500-year-old Machrie Moor Standing Stones, set amid isolated moorland, and to sample a more earthly spirit, head to Lochranza for a whisky-tasting and distillery tour.

Next, take the ferry from Lochranza to Claonaig followed by the short road trip across Kintyre (and yes, it's OK to hum 'Mull of Kintyre' to yourself on the way). From Kennacraig, there's a ferry over the Sound of Jura to the southernmost of the Inner Hebrides, Islay. You're now even deeper in whisky country: to investigate, visit Bowmore, Scotland's oldest (legal) distillery, established in 1779. For a sobering overview of the island's austere past, pop into the Museum of Islay Life, then stretch your legs at RSPB Scotland Loch Gruinart, a reserve that protects redshanks, curlews, corncrakes and butterflies. Come autumn, migratory white-fronted and barnacle geese arrive from Greenland for the winter.

Best time of year: April to October. Peak season is July and August. In summer, midges can be a problem in low-lying areas.

Plan your trip: Allow at least seven days to island-hop by CalMac ferry from Ardrossan to Mallaig. Each leg takes between 30min and 2hr 30min. The ferries carry cars, bikes and foot passengers. ChargePlace public EV charging points are found on all the islands except Jura and tiny Iona. Bringing a car saves time, but with a bit of timetable-crunching, you could use local buses to explore instead.

Getting there: Ardrossan and Mallaig are around 700km (430 miles) and 900km (560 miles) respectively by road from London. By train, travel to Glasgow for rail or road connections to Ardrossan, or to Fort William for Mallaig. Both Glasgow and Fort William are on the Caledonian Sleeper route from London Euston.

The Paps of Jura, a trio of mountains on the island of Jura, are a popular target for hillwalkers.

Islay's wild neighbour, Jura, is just ten minutes away by ferry. You'll get a sense of its big, open spaces if you hike up Beinn an Oir, the Mountain of Gold, a bald hill with a 785-m (2,575-ft) peak. The famous Jura Distillery at Craighouse welcomes visitors for tours, tastings or a friendly chat about island life, past and present.

Nip back to Islay to catch the ferry to Colonsay, an island that's truly off the beaten track. Even though it's just 13km (8 miles) long, there's plenty to do, from beachcombing and golf to bird-watching boat trips.

Over the sea to Skye

From Colonsay, travel northeast to Oban and change ferries for the crossing to Mull. On the way, you'll pass Duart Castle, a lonely clifftop fortress with sweeping views over the Sound of Mull.

On Mull, keep your eyes peeled for white-tailed sea eagles, which have been successfully reintroduced to the Inner Hebrides. For your best chance of seeing some, join Mull Eagle Watch on a guided walk to the Glen Seilisdeir bird hide in Tiroran Forest. Further west, off the island's remotest tip, is Iona, whose 6th-century abbey is a place of pilgrimage; you can visit by passenger boat.

From colourful Tobermory, Mull's capital, take the ferry to Kilchoan and travel up the strikingly picturesque Lochaber coast to Mallaig for the ferry to Armadale on Skye. Rightly one of Scotland's top places to visit, this is an island that deserves to be explored at leisure, drinking in the superb scenery that surrounds its winding roads. There are clear pools to dip in, bays to sail in, mountain trails to hike or cycle along – and on a winter visit, if you're very lucky, you could see something that outshines the rainbows: the northern lights.

ISLE OF STAGS

The name Jura is believed to be derived from *hjǫrtr*, the Old Norse for stag. Today, the island's free-roaming red deer outnumber its two hundred or so human residents by around thirty to one. Other wildlife species that are easy to see here include otters, seals, terns and gannets.

BEYOND THE WHISKY TRAIL

Colonsay is the smallest island in the world with a commercial brewery. It produces three craft beers and has recently tapped into the popularity of botanical spirits, setting up a gin distillery that uses locally foraged plants in its recipes.

The Orkney Islands, Scotland: To visit this outdoorsy archipelago, take the Caledonian Sleeper train: either to Aberdeen for the ferry to Kirkwall, or to Inverness (followed by a bus up the coast to Scrabster) for the shorter crossing to Stromness. Alternatively, drive north by electric car: Orkney has over thirty EV charging points, handy for an eco-friendly road trip. Electric campervans are available to hire on Mainland, the largest island.

Skomer, Skokholm and Bardsey, Wales: These islands are prime bird-watching spots. Passenger ferries and simple visitor accommodation are available in summer. They're popular, so book in advance.

The climb to the Old Man of Storr on the Isle of Skye is one of Scotland's most iconic hikes.

Recreate your childhood summers in Scilly

SOUTHWEST ENGLAND, UK

For old-fashioned, back-to-nature beach holidays, Cornwall's pretty archipelago is just the ticket

Sandcastles whacked with a spade. Dolphins leaping, far out in the blue. Shells, rockpools and soft, gentle waves. Fish and chips wrapped in paper. Ice cream that melts down your chin. If innocent joys like these say 'summer holiday' to you and your kids, the Isles of Scilly might be your kind of place.

Warmed by the Gulf Stream, Cornwall's archipelago has a subtropical climate with over 1,750 hours of sunshine per year, making it ideal for outdoorsy breaks. This cluster of islands is the UK's smallest Area of Outstanding Natural Beauty, with clean sandy beaches backed by bracken and heather. It's so peaceful and safe here, that you can let the kids do their own thing, whether it's paddling, beachcombing or reading. That way, everyone has time to relax and recharge.

Of the five inhabited islands, St Mary's is the largest, with the most accommodation. Its capital, Hugh Town, is tiny – strict planning regulations sanctioned by the Duchy of Cornwall have kept development in check. Many hotels and B&Bs have family rooms, but to really stretch out, book a self-catering pad – a traditional Scillonian granite cottage with cheerful clumps of blue and white agapanthus outside, perhaps, or a barn conversion with sea views. The resort island of Tresco has particularly strong eco-credentials with dedicated recycling, community beach cleans, drinking water points and a general store that has been plastic-free for years. It also has a gorgeous spa and indoor pool.

THE LOWDOWN

Best time of year:
April to September. During school holidays and half-term holidays, accommodation tends to be booked well in advance.

Plan your trip: Allow anything from a long weekend to a fortnight. The islands welcome independent travellers. It's also possible to book organized holidays including accommodation and guided visits.

Getting there:
The *Scillonian* passenger ferry sails from Penzance in Cornwall to St Mary's daily between mid-March and early November, taking 2hr 45min. Penzance is around 460km (290 miles) by road from London, taking around 8hr 20min by bus. Trains from London Paddington to Penzance take around five hours (or 9hr 20min if you opt for the Night Riviera sleeper).

Sailing boats, ideal for exploring Scilly, moored in the channel west of the island of Tresco.

Phoenix Craft Studios on St Mary's was set up by a cooperative of artisans who create prints, ceramics, jewellery and more. Some even offer classes. You'll also find play centres in the islands' community halls. So that's your rainy-day plan sorted.

IN THE KNOW

Scilly's month-long food festival, held in September, celebrates delicious local ingredients and the islands' Slow Food culture. Kids can get stuck into pasty-crimping workshops, cream teas and beach barbecues, while for the grown-ups, Spirit Scilly runs a pop-up gin school, with sixty botanicals to sample.

Active adventures in the Isles of Scilly

Visitors to the Isles of Scilly can't bring vehicles, hire cars aren't available, and the off-islands (those beyond St Mary's) are almost entirely car-free. But that's no reason to stay put. It's easy to plan a whole fortnight's worth of adventures involving walking, cycling or catching the Community Bus, which trundles around St Mary's every day in summer and will stop if you hail it. To island-hop, use the small, brightly painted, open deck ferries. Operated by local boatmen's associations, most were built in the 1930s and 1940s and have jaunty names such as *Seahorse*, *Guiding Star* and *Surprise*.

When you're ready for a break from the beach, visit Tresco's Abbey Garden, which is crammed with Mediterranean plants and has some surprisingly colourful wildlife: red squirrels and golden pheasants. Entry is free for under-fives.

Budding naturalists will love the guided coastal walks and boat trips run each week by the Isles of Scilly Wildlife Trust and St Agnes Boating. Puffins nest on the uninhabited islands in summer, and you can watch the birds coming and going from a boat. Britain's largest carnivores, Atlantic grey seal, breed here and are easy to see – you just need to know where to look.

Below: created in the 19th century among the ruins of a Benedictine abbey, Tresco Abbey Garden has an abundance of palms, proteas and strelitzias. Right: at low tide, boats rest on the shore of St Mary's Harbour.

ALSO TRY

Isle of Wight: Off the Hampshire coast, this is another old-fashioned holiday island that's having a comeback. Getting there by ferry is in itself a scenic experience. There are up to two hundred crossings from Portsmouth, Southampton and Lymington per day.

Isle of Portland: Connected to Dorset's Jurassic Coast by a sliver of sand, Portland has a youth hostel with bell tents, plus a butterfly reserve, bird observatory and child-friendly swimming spots.

Discover the hidden beaches of Finistère

NORTHWEST FRANCE

THE LOWDOWN

Best time of year:
June to September.
High season is July
and August.

Plan your trip: Allow
at least a long weekend,
perhaps as part of a
tour of northern France,
staying at campsites,
gîtes or chambres
d'hôtes.

Getting there: The
village of Morgat on
Finistère's Crozon
peninsula is around
720km (570 miles)
by road and sea from
London via Kent. You
could travel by Eurostar
train from London St
Pancras to Paris Nord,
then TGV train from
Paris Montparnasse to
Brest and bus or taxi
to Morgat (from 8hr
20min). Other options
include travelling to
Plymouth for the ferry
to Roscoff (5hr 30min
by day; 11hr 15min
overnight). Roscoff is
only 90km (55 miles)
from Morgat, taking
1hr 15min by bus. Or,
take the ferry from
Portsmouth to St Malo,
then cross Brittany by
road or train.

Flung out like a toddler's arm into the Celtic Sea, Brittany is as distinctive as an island

With closer cultural connections to Cornwall, Ireland and Galicia than to Paris, Brittany has its own language, gastronomy, music and mystical traditions. Its remote rural landscapes are dotted with gifts from the past – stone cairns, princely tombs and granite menhirs, just like the ones that cartoon hero Obelix, of Asterix and Obelix fame, used to carry on his back.

On the coastal fringes, ancient history and natural history intertwine. Brittany's beaches, dunelands and wetlands harbour an unusual assortment of flora and fauna; otters, beavers, puffins and Montagu's harriers all thrive on the furthest-flung islets, as do carnivorous sundews.

Asterix and Obelix's old hunting grounds lay within what is now the French department of Finistère, a name that neatly echoes Cornwall's Land's End and Galicia's Cabo Fisterra. Here, at Brittany's western extreme, deer, wolves and wild boar used to roam free. The Domaine de Menez Meur, a rural estate that's open to the public, has a collection of native wildlife that includes these species, along with local domestic breeds such as Bretonne Pie Noir cows, Ouessant sheep and West French White pigs.

The cliffs above la plage de l'île on Brittany's Crozon peninsula are tufted with heather, thrift and gorse.

SENTIERS MARITIMES

Created in the 18th century as anti-smuggling patrol routes for customs officers, Brittany's rugged cliff paths – known as *sentiers maritimes* – now attract walkers. Organized walking tours provide daily itineraries, accommodation and luggage transfers that enable you to spend several days exploring.

GO GREEN

Brittany has a good selection of ecologically sound places to stay, from rural campsites, traditional stone cottages and farmhouse gîtes to a family-friendly woodland glamping retreat with bell tents and rustic-chic timber cabins.

Over 2,000km (1,250 miles) long, the GR 34 hiking route traces Brittany's beautiful coastline from Mont-Saint-Michel to Saint-Nazaire, hugging the shores of Finistère on the way.

Wild beaches and secret spots

Finistère's loveliest end-of-the-world landscapes can be found within the Parc naturel regional d'Armorique. Tucked beneath the heather-clad cliffs of the Crozon peninsula, south of Morgat, is la Plage de L'île, a gorgeous little south-facing beach of pebbles and sand, lapped by the calm waters of Douarnenez Bay. Regularly voted France's favourite secret beach, it's pale enough for the sea to look as blue as a swimming pool on bright, summery days. The descent, via a steep, uneven path, is tricky, but the reward is a sun-worshipping spot as beautiful as any in Croatia or Greece.

This is by no means the only special beach in the park. Plage de Lostmarc'h, on the opposite side of the peninsula, facing the Celtic Sea, is windier, wilder and equally undeveloped. In the clothing-optional southern section, World War II anti-landing defences still poke out of the sand. Come here to surf, if you're bold and experienced (the currents can be challenging, so swimming is not recommended), or to enjoy a proper leg-stretch. Backed by grassy coastal heathland, the broad, beautiful sands are 2.5km (1.5 miles) long.

The beach at Morgat is excellent in a different way. Since it's a town beach, it can get busy, which, in this refreshingly spacious corner of France, can come as a culture shock. The upside is that once you've worked up an appetite, you can quickly retreat into a bistro to feast on moreish crêpes, garlicky *fruits de mer*, tangy *charcuterie* and creamy, squidgy cheese. The local cider is sure to lift your spirits, too.

ALSO TRY

Guernsey, Channel Islands: The cliff paths that wind along the southeastern coast offer thrilling sea views. Steps lead down to Moulin Huet, Le Jaonnet Bay and other hidden coves, for secluded swimming.

Ceredigion, Wales: The underrated coast between Cardigan and Aberaeron is wonderful for walking, with remote, secluded National Trust beaches such as Mwnt and Penbryn that only the locals know.

Finistère's more unusual places to stay include a solar-powered former lighthouse keeper's cottage on Île Louët, a rocky islet in Morlaix Bay near Carantec, southeast of Roscoff.

Escape to Formentera's endless sands

BALEARIC ISLANDS, SPAIN

Ibiza's tranquil little sister has dazzling white Mediterranean beaches on every coast

Dangling like a jewel from Ibiza's southern lobe, Formentera is constantly compared to its larger, louder neighbour. That suits its admirers just fine. In a great many respects, it compares very well indeed.

Where Ibiza has overdeveloped resorts, Formentera has get-away-from-it-all villages. Lovers of rural Ibiza will point to its whitewashed fincas and picturesque pines, but they're nothing compared to the natural grace of Formentera's long, lovely, thinly developed beaches. And the atmosphere is substantially different; while Ibiza draws a frenetic crowd of hard-partying clubbers, promoters and influencers, Formentera attracts a laidback assortment of modern-day lotus-eaters and Euro-bohemians, many of them free-spirited naturists, artists and musicians. Yes, there are cafés, bars and clubs here – it's definitely not dull – but the mood is more discerning and, dare we say it, grown-up.

Only around 20km (12 miles) long, Formentera has a single low-rise beach resort, Es Pujols, with a pretty promenade, an appealing beach and a watersports centre, where you can hire kayaks, windsurfers, stand-up paddle boards and Topcat catamarans.

But that's just the start. With warm, shallow water and perfectly soft sand, beaches like the 8-km- (5-mile)-long Platja de Migjorn on the south coast and Platja de Llevant on the Es Trocodores peninsula are the type that escapists adore – particularly those who prefer their sun-worshipping to be clothing-optional. Best of all, Formentera's precious coastal lagoons and seagrass beds, protected by Ramsar and UNESCO, help keep the waters pristine.

FOLLOW THE GREEN ROUTES

Formentera is small enough to explore on foot or by rented bicycle. To help you discover its hidden corners, pick up or download Formentera Tourism's guide to the island's 32 Green Routes, which lead through dunes, pine groves and traditional hamlets. Electric scooters and electric cars are also available to hire and are easy to charge.

THE LOWDOWN

Best time of year: April to October. From November to March, there are fewer ferry crossings. High season is July and August.

Plan your trip: Allow at least a couple of days. Some visitors stay all summer.

Getting there: The fastest ferry crossing from mainland Spain to Formentera departs from Dénia in Alicante, taking 2hr 30min. Dénia is around 1,940km (1,200 miles) by road and sea from London via Kent, France and northeast Spain. Stringing together trains from London St Pancras via Paris and Barcelona to Valencia with the bus to Dénia and the ferry to Formentera takes 20 hours, excluding stopovers, so it's best to allow two days. There are also longer crossings from Barcelona and Valencia.

ALSO TRY

Île de Ré, France:
Parisians would love to
keep this sleepy spot in
the Bay of Biscay secret.
It has pretty villages
and beaches backed
by dunes and pines.

Sylt, Germany: At
Schleswig-Holstein's
northernmost tip, this
island has long sands
with distinctive stripey
chairs called *Strandkörbe*
(beach baskets).

Ride the waves in Ericeira

WEST PORTUGAL

With forty superb Atlantic beaches, this breezy stretch of coast is Europe's first World Surfing Reserve

The waves that crash onto Ericeira's sandy beaches are remarkable for their power and variety. The surf zone contains no fewer than seven famous breaks, known as Pedra Branca, Reef, Ribeira d'Ilhas, Cave, Crazy Left, Coxos and São Lourenço. While beginners can have a stab at Ribeira d'Ilhas, others, such as Coxos, are gnarly enough to challenge champions.

Surfers care a great deal about the health of our oceans. You could say they're on the frontline. It's no coincidence that one of Britain's leading marine conservation organizations was founded by surfers, calling themselves Surfers Against Sewage. These days, Surfers Against Sewage And Single-Use Plastic would be even more apt.

The California-based conservationists, Save The Waves, were so impressed by the 4-km (2 1/2-mile) Ericeira coast that in 2011, they declared it Europe's first (and, so far, only) World Surfing Reserve (WSR), the second in the world after Malibu in California. Eight more destinations including Australia's Gold Coast and Chile's Punta de Lobos have since joined the list.

The WSR programme seeks to protect surf breaks of outstanding environmental, economic and cultural value. Ericeira is special because it's a marine biodiversity hotspot whose local community, the Jagoz, live entirely from the sea. In previous decades, they fished for sardines, tuna and mackerel, sailing out into the Atlantic and surfing their boats home. More recently, they've set up a string of surf schools and shops, and are doing a roaring trade.

THE LOWDOWN

Best time of year: May to October. For beginners, the early part of the season is best. High season is July and August. Leave the end of the year (September to January) to the experts: winter conditions can be cold and challenging.

Plan your trip: Allow at least a long weekend in Ericeira (pronounced *airi-sigh-ra*), perhaps as part of a tour of Portugal, Spain and Morocco.

Getting there: Ericeira is around 2,205km (1,370 miles) by road and sea from London via Kent and France, and 48km (30 miles) by road from Lisbon. By train from London, you could travel via Paris to Bayonne or Hendaye in the Basque Country (8hr 30min), then continue by train via Madrid to Lisbon (19hr), or catch the overnight bus from Bayonne to Lisbon (13hr 30min). Alternatively, take a ferry from Plymouth or Portsmouth to Santander or Bilbao (18–24hr), followed by trains via Madrid to Lisbon (17hr 30min).

IN THE KNOW

Like all the world's best surf spots, the town of Ericeira is beguilingly laid back. Be sure to wander its cobbled streets and squares, stopping at cafés for a *bica* (Portugal's answer to espresso) and a crunchy and creamy *pastel de nata*.

ALSO TRY

Rossnowlagh, Republic of Ireland:
On Donegal's Atlantic coast, this is a top surfing destination for beginners and improvers, and the après-surf *craic* is top notch.

Hossegor, France:
This surfing fanatic's favourite, north of Biarritz, hosts international championships including Quiksilver Pro France.

Island-hop from the Côte d'Azur to Calabria

SOUTH FRANCE, ITALY AND MALTA

Cross the Ligurian and Tyrrhenian seas by ferry to explore the Mediterranean's largest islands and one of Europe's tiniest nations

Bordering one of the wealthiest coastal areas in Europe, the Ligurian Sea is rich in natural history, too. As you bid au revoir to the chic, sun-dappled boulevards of the Côte d'Azur and board the ferry to Corsica, you enter the domain of whales and dolphins.

The entire coast and sea between Toulon, Tuscany and Sardinia, an area of 87,500km² (33,784 square miles), is a protected area, the Pelagos Sanctuary. Its main purpose is to safeguard marine mammals against boat traffic, fishing nets, chemical pollution, underwater noise and other disturbances. Eight cetacean species are present throughout the year, including between 20,000 and 45,000 striped dolphins. Sociable and acrobatic, they often leap alongside the bows of boats, so it's worth keeping watch whenever you're afloat.

Between 2,000 and 10,000 long-finned pilot whales, the largest oceanic dolphins after orcas, are also found here, and more than 1,000 fin whales swim in the deepest waters, 1,000m (3,280ft) down. Both appear in summer, when food is abundant. Bottlenose dolphins, which feed in the Tyrrhenian Sea, close to the Corsican and Sardinian coasts, can be seen all year round.

The possibility of spotting cetaceans adds magic to your travels as you take to the water. Daytime ferries take between four and ten hours to cross from the Côte d'Azur to Corsica: ample time to relax and train your eyes. Crossings from Corsica to Sardinia can take less that an hour. The weekly voyage from Sardinia to Sicily in the big one, taking 12 hours, while the trip from Sicily to Malta is a little under two.

Across Corsica, on foot and by train

Many visitors choose the Mediterranean islands for their sunshine and wine. But if you're seeking something more challenging, they also deliver. Corsica has one of Europe's toughest Grande Randonnée long-distance paths, the GR20. Hiking for six and a half to eight hours per day, you'll need two weeks to complete it, and legs of steel.

If that sounds a little extreme, you could explore by train instead, enjoying the island's best sea views and spectacular rocky interior aboard the Trinighellu, a historic narrow-gauge railway. Founded in the 1850s, this jaunty little boneshaker runs between the island's four main ports, Calvi, L'Île Rousse, Bastia and Ajaccio, via Vizzavona and Corte in the centre.

Perched on the limestone cliffs of southern Corsica, the medieval town of Bonifacio has a superbly dramatic setting, close to uncrowded sandy beaches.

THE LOWDOWN

Best time of year: April to October. High season is July and August. Some ferry services are reduced from September or October to May or June.

Plan your trip: Allow at least ten days to visit Sardinia, Corsica, Sicily and Malta. Each island has several ports, with numerous ferry companies operating to and between them.

Getting there: Ferries from the Côte d'Azur to Corsica leave from Marseille, Toulon and Nice. Marseille is around 1,240km (770 miles) by road and sea from London via Kent. Travelling by rail from London St Pancras takes around 6hr 30min via Paris (transferring from Paris Nord to Gare de Lyon via Transilien) or 7hr 30min via Lille (without changing stations). Sicily and Calabria are around 21–25hr from London by train. InterCity and sleeper trains between Sicily and Naples, Rome and Milan cross the Strait of Messina by train-ferry – a unique experience.

Cycling in Sardinia

The Mediterranean islands' clement weather is ideal for cycling. Corsica's hill-climbs suit would-be pros in Tour-de-France-style lycra. In contrast, neighbouring Sardinia, with its flat coastal roads, is more benign, with broader appeal. Don't mistake benign for boring, though. To crank the experience up a notch, you just need to venture inland, where, particularly in the north, there's tougher terrain to tackle.

When planning a route it's worth clocking the fact that Sardinia, at around 24,000km² (9,266 square miles), is larger than Wales but less than half as populated. As long as you're fairly self-sufficient, its quiet, traffic-free roads – leading through vineyards, olive trees and cool, shady forests – lend themselves to long-distance rides.

Staying on a Sicilian lemon farm

If the idea of Sicily conjures romantic thoughts of scented lemon groves, you'll love the island's *agriturismos* (farmstays). With guest rooms in converted farm buildings or stone-built farmhouses, they offer a peaceful, rural experience. Once you're settled in, you can wander through the orchards on foot, or venture further afield by bicycle.

As well as having dramatic beaches, ancient Greek temples and Roman mosaics to discover, Sicily is also a place to indulge in culinary delicacies. Expect endless temptation: the islanders make excellent *torroncini* (soft nougat), *paste di mandorla* (almond biscuits), *vino alla mandorla* (almond wine) and that Italian favourite, *limoncello* (lemon liqueur) served ice cold.

ECOTOURISM

Monitor whales and dolphins: Between May and September, Tethys, a nonprofit research organization, invites volunteers to participate in their whale and dolphin research programmes in the Ligurian Sea (Italy) and the Ionian Sea (Greece). Opt for the Italian project, and you'll spend six days living at sea on an expedition yacht, helping collect and analyse data in the Pelagos Sanctuary.

Holiday accommodation with heart: Eco-friendly places to stay are popping up on the Mediterranean islands, from Cocoon Village on Corsica – a cluster of fabric bubbles perched among fragrant conifers on a cliff face – to Sardinna Antiga in the heart of Sardinia, a bio-sustainable hamlet created by recovering ancient, abandoned *pinnatu*, shepherds' huts.

On Sicily, Taormina's Teatro Antico, a Greco-Roman amphitheatre with splendid views of the Calabrian and Sicilian coasts and Mount Etna, is still in use today, hosting theatre, cinema and opera galas.

Scuba diving in Malta

Sicily is so close to Calabria – the toe of the Italian boot – that it feels as if you could swim there: on special race days, top athletes make it across the Strait of Messina in around thirty minutes. But before you jump on a ferry or sleeper train to the mainland, tiny Malta, 100km (62 miles) south of Sicily's southern coast, makes a worthwhile diversion. It's not one island, but three. Northwest of the mainland is Comino, much of which is a bird sanctuary where yelkouan shearwaters breed in summer; electric boats ferry passengers to its Blue Lagoon (*pictured, right*) for swimming, snorkelling and strolling. Beyond lies Gozo, one of the Mediterranean's best scuba diving spots.

Gozo's wildlife-rich underwater landscapes are truly impressive, and calm seas ensure excellent visibility. The best reefs and wrecks are dotted all around the island, within easy reach, so if conditions are a little choppy at one site, you can easily travel to another. Good places to start include Marsalforn's Santa Maria Breakwater on the north coast, Dwejra's Blue Hole to the west, and, in the south, Xlendi Tunnel, Ta' Cenc Cave and Fessej Rock near Mġarr ix-Xini.

Above: the Blue Grotto, a complex of sea caves on Malta's south coast, is a popular destination for snorkelling trips, setting out from Wied iż-Żurrieq harbour. Right: traditional Maltese fishing boats dot the bay at Marsaxxlok.

SAY HELLO IN MALTESE

Maltese has an Arabic twang. It pulls together words from Arabic, Italian, English and French, and is the only Semitic language to be an official language of the European Union. To break the ice, say, 'Hello, bonġu' (pronounced bong-ju), which means, 'Hello, good morning'.

ALSO TRY

Italy's smallest islands: Giglio and Elba in the Tuscan archipelago, the Pelagie Islands south of Sicily, the Tremitis in the Adriatic, Ischia and Procida in the Bay of Naples and Ponza south of Rome are all popular with locals and reachable by ferry in high season.

Hidden-gem Greek Islands: Why settle for Santorini or Mykonos when Greece has so much more to offer? Underrated spots include Milos and Koufonisia in the Cyclades, Elafonisos in the Peloponnese and Hydra in the Saronic Islands.

Island-hopping in the Canary Islands: Rich in volcanic landscapes, this North African archipelago has distinctive cultural traditions that set it apart from the Spanish mainland, 1,400km (870 miles) away. Ferries make it surprisingly easy to reach and explore.

Unwind on a Mediterranean gulet

TURKEY AND GREECE

A gulet cruise is all about relaxation, exploring islands and bays that larger boats can't reach

Gulet cruising is sailing for softies. You'll be living on a yacht with comfortable cabins, ample deck space and a skipper, chef and crew who are dedicated to making everything run smoothly. All you have to do is show up at the starting point – such as Bodrum on Turkey's Turquoise Coast, or the Greek island of Rhodes – and relax with your friends or family as the glorious indigo sea rolls past. It's much like a floating house party.

From time to time you'll go ashore, avoiding the tourist traps that teem with holidaymakers every summer. Free to roam, gulets drop anchor in hidden coves, and send you by tender or kayak to beaches where your footprints in the sand are the first of the day. Sometimes, you might visit archaeological sites, or wander through quiet fishing villages, pulling up chairs at a taverna for lunch or a glass of wine with the locals.

Sailing the Med in style

Southwest Turkey is the birthplace of the gulet. Traditionally, these elegant wooden schooners were used for fishing, sponge fishing and freight in the Mediterranean, Aegean and Black Seas. These days, most are skippered charter vessels, launching from marinas in Italy, Montenegro and Croatia as well as Turkey and Greece.

It takes the master boatbuilders of Bodrum and Marmaris between nine months and two years to construct a gulet from scratch. Built of white, red and black Aegean pine, they have two or three masts and are between 15–36m (50–120ft) in length, with four to eight cabins.

Gulets (and caiques, their Greek equivalent) vary a great deal in the level of luxury they offer. At the top end, you can expect first-class treatment and extras such as aircon, a jacuzzi, watersports gear and large en-suite shower rooms.

THE LOWDOWN

Best time of year: May to June or September to October. Avoid July and August, when conditions can be hot and windless, and the Aegean beach resorts tend to be crowded.

Plan your trip: A typical gulet cruise lasts eight nights and follows a pre-planned itinerary from Bodrum in Turkey or Rhodes in Greece. Some operators offer bespoke trips: exploring the Greek Islands from Kos or Corfu, for example.

Getting there: Bodrum is around 3,750km (2,300 miles) by road and sea from London via Kent, Germany and the Balkans; Rhodes is 100km (60 miles) further south. There's no high-speed rail route from the UK and travelling by train and bus takes well over two days. You may prefer a three-day (or longer) adventure by train and ferry: whizzing down Italy to Bari or Brindisi by train, sailing to Patras in Greece, then continuing by bus and train to Piraeus near Athens for a ferry across the Aegean.

GULETS FOR GOURMETS

Gulet chefs work wonders, rustling up fabulous Turkish- or Greek-style meals. Dinner might start with meze such as olives, oil-drizzled salads and fried vegetables dressed in yogurt, followed by fresh local sea bream, swordfish or lamb, all served under the stars.

IN THE KNOW

Kumbahçe – Bodrum's southeastern quarter – is the heart of Turkey's gulet-building industry. Wander along Içmeler Caddesi (Içmeler Street, named after the resort and bay on the Datça peninsula) and Gület Caddesi, to hear and see craftsmen at work.

Crucially, despite having masts and rigging, some gulets don't sail; instead, they travel by engine power, to allow the crew to focus on service. This isn't adrenaline motoring – the captain will probably only run the engine for two to four hours a day, and will rarely top 10 knots (11.5mph) – but it's not unusual to burn over 1,500 litres (330 gallons) of fossil fuel in an eight-night voyage. If you'd like your trip to be as eco-friendly (and peaceful) as possible, you could request a gulet that's rigged and crewed for sailing. Bear in mind, though, that a 36m (120ft) gulet needs a great deal more wind power than, say, a lightweight 12m (40ft) yacht, and the skipper will lower the sails when it's simply too calm.

If a chance to turn off the engine arises, you'll enjoy an immediate change in atmosphere. There's an ancient magic to the sound and feel of a boat that's powered by sail. If you're interested in helping out, the crew will be glad to show you the ropes. Just don't expect them to have quite as much time for pouring champagne as when you're motoring.

Treasures from the deep are on display in the underwater archaeology museum housed in Bodrum's 15th century castle.

ALSO TRY

Montenegro: Explore the coast on a gulet loaded with mountain bikes, heading to shore for adventurous cycle rides by day and spending your nights on the boat.

North Atlantic: Join a skippered sailing trip from Scotland to the Orkney, Shetland and Faroe islands, then on to Iceland and Norway's Lofoten Islands.

Whitsunday Islands, Australia: Charter a skippered sailing yacht in this beguiling sailing destination. Rich in wildlife, it's within easy reach of Airlie Beach in Queensland.

Once a hangout for smugglers and pirates, Ölüdeniz, with its indigo lagoon and Blue Flag beach, has become one of southwest Turkey's most appealing gulet cruise destinations.

Kayak your way along the Dalmatian Coast

SOUTH CROATIA

Best time of year: May to October. High season is July and August. In September, the weather is perfect. Some ferries have a reduced service from October to May.

Plan your trip: Allow five to fourteen days. Group holidays are typically seven to eleven days. Starting points include Zadar, Split and Dubrovnik.

Getting there: The Dalmatian Coast is around 1,300km (800 miles) by road and sea from London via Kent, Germany and Austria. Trains from London St Pancras via Paris and Munich to Zagreb take 25 hours. From Zagreb, take a bus to Zadar (2hr 30min) or the superbly scenic train to Split (6hr). In summer, a daily ferry connects Split to Dubrovnik via the islands of Brač, Korčula and Hvar. There are also buses and private minibus transfers along the coast.

You could travel the Croatian Adriatic by ferry, dinghy or yacht, but there's nothing quite like venturing out under paddle power

Croatia is blessed with some of Europe's loveliest islands. The Dalmatian Coast between Istria and Dubrovnik is strewn with them: slim shards of sand and rock, tufted with Aleppo pines, holly oaks, laurels and cypress. Most lie at a northwest-southeast angle, shaped by tectonic activity in the Mesozoic era and separated from the mainland by channels that flooded around 11,000 years ago.

There's really no reason to linger in Dubrovnik's beautiful but overcrowded Old Town when waterborne adventures await. All told, the Dalmatian archipelagos include 79 islands and around 500 islets. Some have pretty fishing villages of stone-built, red-roofed houses, set beneath hills planted with vineyards and olive groves. A few, including Hvar and Brač, are blessed with eco-friendly holiday villas with solar power, rainwater harvesting and organic gardens. These idyllic islands lie in a glorious, glossy stretch of the Adriatic, within 120km (75 miles) of the coast. Most are considerably closer than this, with short distances between them, making long-distance island-hopping by sea kayak an attractive and thoroughly feasible prospect.

Beyond an ability to swim, you don't need much in the way of natural know-how or training; if you book an organized trip, your guide will assist with the latter. You can tailor your itinerary to suit your energy levels, paddling gently for an hour or so at a time if you'd just like a taster, or spending five hours or more per day on the water, investigating natural arches, inlets and caves, if you'd prefer a proper expedition.

The Elaphiti Islands, immediately northwest of Dubrovnik, are a good target for beginners. Alternatively, further north, there's the Zadar archipelago, with its pine-scented islands and crystal-clear water, perfect for snorkelling, and the Kornati archipelago, a national park of beaches, rocky headlands and dry-walled pastures.

Soon you'll be confident enough to make your own way from one island to the next, either leaving your luggage to be transferred by boat or, in the interests of saving fuel, carrying it with you – assuming you've packed supremely light. Wherever you go, you'll be dazzled by the Adriatic's huge skies and vivid blue hues.

ISLAND OF FESTIVALS

The popular Dalmatian island of Hvar has a packed calendar of events, with celebrations of wine, lavender and music lasting all summer. One of the best gatherings is the five-day *Dani u Vali* (Days in the Bay) festival, held in September in the harbour village of Stari Grad. It features boat races, culinary tastings, outdoor concerts and sailors dressed up as pirates.

Spotted dogs and soparnik

On dry land, you'll have plenty of time to chat to the locals and learn about the region. Let's get the obvious questions out of the way: yes, Dalmatian dogs did probably originate in Dalmatia; they're mentioned and depicted in Croatian church chronicles, paintings and frescoes dating back to the 16th century. Possibly more worth knowing, however, is that the islanders' olive-oil-rich, farm-to-fork Mediterranean diet is such a powerful facet of community life that it features on the UNESCO Intangible Cultural Heritage list. With home-cooked food shared among families and neighbours, every meal is an enthusiastic and convivial affair – all the more reason to grab another slice of *soparnik*, a flat, crispy pie stuffed with chard, or tuck into a fresh plateful of *crni rizot* (squid-ink risotto) with garlicky shellfish, oil-drizzled greens and fruity Plavac Mali wine.

ALSO TRY

Byron Bay, Australia: This is easily one of the most appealing coastal resorts in New South Wales. Take to the water in a kayak, and there's a good chance you'll soon have bottlenose dolphins for company.

Aegean Sea, Greece: Paddle around Crete, or make your way from island to island, staying overnight in friendly tavernas with guest rooms.

Shetland Islands, Scotland: Sea kayak along the archipelago's rugged coastlines, exploring huge, complex sea caves lit by shafts of sunlight.

Choose your island in the Stockholm Archipelago

SOUTHEAST SWEDEN

Lose yourself in a labyrinth of islands lapped by clear, calm water, a ferry ride away from Sweden's eco-friendly capital

The locals call it *skärgård*: garden of islets. Fanning out some 60-80km (37-50 miles) from the Swedish capital, the Stockholm Archipelago – a maze of 30,000 Baltic islands, skerries and rocks – is Sweden's largest group of islands.

While Swedes think of the archipelago as an escape from the bustle of city life, the capital is actually right in its midst, and the islands are not all wildernesses: two hundred of them are inhabited, linked by a network of three hundred mainland and island ports and jetties.

The town of Vaxholm, one of the most popular destinations, has a population of just under 5,000 and a road bridge to the capital, while Sandhamn, another locals' favourite, has a busy little harbour village and marina; both are well under two hours from Stockholm by ferry.

But for the most part, it's the archipelago's blue waters, quiet beaches and whispering pines that are the main draw. Over forty of the islands are nature reserves owned by Skärgårdsstiftelsen (the Archipelago Foundation), including peaceful Grinda, which has child-friendly sandy beaches and an ice-cream kiosk, and Finnhamn, known for its beautiful deciduous woodland and organic farm shop. Möja may not have swimming beaches, but its pathways beg to be explored on foot or by bike. All three have low-key accommodation options including hostels, timber cottages and campsites, several of which have kayaks, rowing boats or stand-up paddle boards for hire.

THE LOWDOWN

Best time of year: April to September. High season is from midsummer in late June to late August. Some services are reduced between October and March.

Plan your trip: Allow two to ten days. A few boat trips are included in the Stockholm Pass sightseeing package.

Getting there: Stockholm is around 1,900km (1,200 miles) by road and sea from London via Kent, Germany and Denmark. From 2021, trains from London St Pancras will connect with a sleeper service from Berlin, and a sleeper from Brussels will launch in 2022, but for now the fastest option, taking 24 hours, involves switching to a night bus from Hamburg. Alternatively, stay overnight in Hamburg, then continue by train the next day. Other routes from the UK include taking a ferry from Harwich, Hull or Newcastle to The Netherlands, then continuing by road or train.

South of Scandinavia, most European languages share variants of the word archipelago. Originally, it wasn't a term for a group of islands: *Archipelago*, meaning 'great open sea' in ancient Greek, was an early name for the Aegean. Since that sea has so many islands, the modern meaning emerged.

IN THE KNOW

The Waxholmsbolaget ferry company sells single-trip, five-day and thirty-day ferry passes on board, and offers an excellent island-hopping map and app with suggested itineraries and things to do.

A journey back to nature

On Svartsö, one of the leafiest islands, visitors can walk or cycle through flower-filled meadows and swim in pristine lakes where beavers are sometimes seen. Bicycles and glamping tents are available and there are cabins to rent, with or without a sauna.

As you'd expect, the further you get from Stockholm, the more remote the atmosphere. The coastal town of Dalarö, southeast of Stockholm, makes a useful base: it's on several ferry routes and has cheaper accommodation than the capital. From here, it's easy to visit sleepy, rural Fjärdlång, less than an hour and a half away via Ornö and a string of other islands, dotted with grazing sheep and red-painted cottages.

How to explore? Gorgeous island scenery comes thick and fast as you travel through the archipelago, making transfers a supremely enjoyable element of any visit. As you'd expect from an eco-conscious nation with a proud maritime tradition, local transport companies have been eager to introduce environmentally friendly hybrid ferries. The first, the *Yxlan*, an ice-breaker based at Furusund, covers the northern part of the archipelago. Elsewhere, it's hard to beat the Waxholmsbolaget steamboats *Västan*, *Norrskär* and *Storskär*, which have old-fashioned character; they serve Vaxholm and the central islands. To hail a ferry, you simply flip the dockside semaphore paddle into the upright position, as a signal to the captain that you'd like to board.

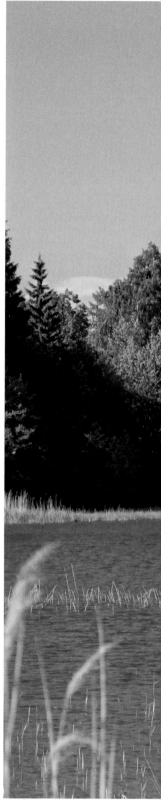

ALSO TRY

Fjord Coast, Norway:
Island-hop around
Norway's thrillingly
remote westernmost
islands including Gåsvær
and Bulandet, travelling
from Bergen by local
ferry and post boat.

Åland Islands, Finland:
Part of the Finnish
Archipelago Sea,
northeast of the
Stockholm archipelago,
these islands are
reached by a scenic
ferry crossing from
Stockholm, Turku or
Helsinki. A beautiful
setting for a back-to-
basics camping or
cottage holiday, they're
veined with cycling
routes and dotted with
quiet beaches and
cosy restaurants.

Ærø, Denmark: Choose
this island for its sleepy
villages, bike-friendly
lanes and simple
country meals. It's
connected to Fynshav
on the mainland by the
world's largest fully
battery-powered car
ferry, Elfærgen *Ellen*
(E-Ferry *Ellen*).

Relax on Essaouira's Plage Tagharte

NORTHWEST MOROCCO

See a different side to Morocco in a refreshingly laidback coastal city

Visiting a Moroccan city can be a frenetic experience, with the cacophony of motorbikes and scooters, the jostling of hawkers and couriers, the vibrant colours of textiles and ceramics and the spicy aromas of busy kitchens and market stalls colliding to the point of sensory overload. Essaouira is different. Here, the madness is dialled down a few notches. Compared to Marrakech, Rabat or Fes, it's positively laid back. And compared to the better-known Atlantic beach resort and party town of Agadir, it's refreshingly bohemian, with rich Tamazight Berber traditions.

Life moves at a gentler pace in this breezy coastal city: locals sit in the doorways of their whitewashed houses, chatting amiably, and even the curio sellers in the medina are more relaxed, happily taking no for an answer.

Essaouira hugs a natural sandy beach, Plage Tagharte, which curves around an island, Mogador. Though only 3km (2 miles) long, it shelters the city from the worst of the Atlantic storms, while funnelling enough breeze to power a modest but buzzing surfing, windsurfing and kitesurfing scene. On bright days, flocks of colourful sails fill the sky.

Board and kite buffs come here to ride the *alizés*: the northeasterly trade winds that blow all summer, making beach picnics a challenge (nobody wants sand in the couscous). The corniche running along the seafront is home to several small outdoor activity centres that offer watersports equipment and training. If you're a beginner, their instructors will show you the ropes on dry land before you tackle the sparkling blue. And when at last your limbs won't take any more, Essaouira's courtyard cafés will beckon you ashore.

FESTIVAL GNAOUA ET MUSIQUES DU MONDE

Essaouira's Gnaoua and World Music Festival is one of Africa's most colourful and joyous musical events. Held in June at numerous venues in the city centre, from intimate courtyards to giant stages on the corniche, it focuses on Gnaoua, the mystical musical style of the Tamazight Berbers, with a dose of African jazz and rock mixed in.

THE LOWDOWN

Best time of year: Any time. For watersports, May to September is best. High season is from late June to late August.

Plan your trip: Allow at least three days in Essaouira, perhaps staying in a converted riad, as part of a tour of Spain, Portugal and Morocco.

Getting there: Essaouira is around 2,500km (1,500 miles) by road and sea from London via Kent, France and Spain. Travelling from the UK by train or bus, allow at least 48 hours, plus stops. Start by making your way to Marrakech via the ferry between Tarifa and Tangier (see page 66), then cover the last 185km (115 miles) to the coast by bus (3hr).

ALSO TRY

Tarifa, Spain: If you're into windsurfing and kitesurfing, it's well worth breaking your journey here before catching the ferry to Africa. Training centres line the Atlantic shore.

Fuerteventura, Canary Islands: Enjoy superb watersports facilities, amid dunes the colour of clotted cream. There are no ferries to the Canaries from Morocco, but overnight services sail from Cádiz and Huelva in Spain.

RAILWAY STORIES

Trains may be getting faster, sleeker and considerably more fuel-efficient, but the loveliest railway journeys remain as romantic as ever. Perhaps it's because they're shared experiences, ripe with potential. In fact, some say a railway carriage is like a library, and its passengers are its books. But what if you just want to gaze out of the window? That's fine too: choose a scenic route through Europe. It's sure to be just the ticket.

Steam across the Harz Mountains

NORTHERN GERMANY

Best time of year:
Any time: trains run
every day of the year.
High season is from
July to October.

Plan your trip: The
Brocken Line from
Wernigerode to Brocken
takes around 1hr 30min.
There's a flat fare,
valid all day, allowing
you to hop on and off;
proceeds help maintain
the entire network. To
enjoy the Harz region
to the full, allow
several days.

Getting there:
Wernigerode in
the district of Harz
is around 940 km
(580 miles) by road
and sea from London
via Kent, Belgium and
northern Germany.
Rail routes from the
UK typically begin
with a leg to London
St Pancras. From here,
trains via Brussels and
Hanover take around
9hr 30min. Alternatively,
travel to Harwich, Hull
or Newcastle for a
ferry to Belgium or the
Netherlands, continuing
to Harz by road or train.

Let the blow of the whistle and the chug of the engine transport you back to a golden age

Twisting and turning through Saxony-Anhalt's steepest, narrowest valleys, the Harzer Schmalspurbahnen (Harz Narrow-Gauge Railway) is a steam buff's dream.

Preserved as a historic monument since 1972, it's Germany's largest network of narrow-gauge railways, with 48 stations, 400 bridges and around 140km (87 miles) of track, set amid the picturesque, forest-draped Harz mountains.

The oldest of the 25 vintage locomotives puffing through this splendid location, dates back to 1897. Others, built in the 1950s, draw traditional open-platform 1930s carriages. Unlike some steam trains, they're not just a tourist attraction – this is an original working railway that has never been mothballed or modernized.

A train ride on the Harz Narrow-Gauge Railway is a genuine highlight of a visit to this part of Germany, perhaps to explore Harz National Park, the UNESCO World Heritage listed town of Quedlinburg or colourful Wernigerode with its castle, museum and Gothic buildings. Enthusiasts can learn more about the railway at the workshop, and can even take a short course to become an honorary driver.

The most popular route in the network is the Brockenbahn (Brocken Line), which branches off the main Harz Line between Wernigerode and Nordhausen and makes a spiralling climb up the Brocken which, at 1,141m (3,743ft), is northern Germany's highest peak. Brocken station, the terminus, is 1,125m (3,691ft) up, just short of the summit; its granite station building, complete with restaurant and souvenir shop, is the highest in Germany to be served by a railway without a rack system to haul it up the slopes.

ALL ABOARD!

Over a million passengers travel on the Harz Narrow-Gauge Railway each year.

ALSO TRY

Saxon Steam Railway Route, Saxony: South of Leipzig, this is a network of vintage railways and museums run by enthusiasts.

Koblenz–Trier Railway, Rhineland: Enjoy a 130-km (81-mile) train journey through picturesque vineyards close to the River Mosel in western Germany.

Spiral through the Alps on the Bernina Express

SWITZERLAND AND NORTHERN ITALY

Feast your eyes on World Heritage listed views as you wind your way from the Swiss Alps to Lombardy by train

Don't let the name fool you: the Bernina Express isn't a particularly rapid means of getting from A to B. But for its passengers, that's no hardship. This famous train caters for sightseers, rather than regular travellers: it takes a leisurely four hours or so to chug, rattle and grind the 144km (89 miles) from Chur, in the beautiful Swiss canton of Graubünden, to Tirano in Lombardy. With gorgeous views of valleys, lakes, mountains and glaciers at every turn and panoramic windows to frame them, the gentle pace is positively welcome. The train also offers modern comforts, from a roving minibar to free Wi-Fi.

The route stitches together two stretches of track: the Albula Line, built in the late 19th century, and the early 20th-century Bernina Line. Together, they comprise the UNESCO World Heritage Rhaetian Railway, boasting 196 bridges, 55 tunnels, superbly engineered spiral loops designed to tackle gradients of up to one in seven, a ride across the Bernina Pass and a 1,800-m (5,905-ft) plunge from Ospizio Bernina to Tirano.

Over the Swiss Alps to Italy

From Chur, the train follows the Rhine then enters the Domleschg Valley, known for its historic ruins and castles. After Thusis on the River Hinterrhein, it continues along the River Albula to Tiefencastel, crossing the River Landwasser via the breathtaking, curved Wiesner Viaduct on the way to Filisur. Next comes the first of the famous spiral tunnels that corkscrew the train up the mountains. In the 5km (3 miles) or so between Bergün/Bravuogn and Preda, the Express climbs an impressive 400m (1,312ft) without the assistance of a rack system.

THE LOWDOWN

Best time of year:
The Bernina Express runs from Chur in Switzerland to Tirano in Italy daily, all year. High season is from June to September. In winter, there are beautiful snowscapes to enjoy.

Plan your trip:
Seat reservations are obligatory. To explore the region around Tirano in more depth, take the connecting bus to Lake Como and Lugano, sitting on the left for the best views.

Getting there:
Chur and Tirano are around 1,060km (660 miles) and 1,240km (770 miles) respectively by road and sea from London via Reims and Zürich. There are trains to Chur from London St Pancras via Paris and Zürich (9hr) and from Tirano to London via Milan and Turin (13hr). Ferries from Harwich, Hull or Newcastle to Belgium or the Netherlands connect via Frankfurt.

WORLD'S NUMBER ONE?

Mark Smith, creator of the railway website The Man In Seat 61, describes the Bernina Express train journey as, 'one of the most scenic in Europe, or for that matter, the world, and a personal favourite too'. High praise indeed.

After the Albula Tunnel, you enter the Bever Valley, part of the Engadin and St Moritz region, famous for winter sports. Both the Albula and Bernina lines connect to St Moritz via a loop of track.

At Pontresina, a mountain village flanked by the Bernina massif, the line changes to a different current, so there's a switch of locomotives. Next, it's a steady climb up to the Bernina Pass at 2,328m (7,638ft) above sea level via Bernina Diavolezza, where you could pause to take a cable car up to the Diavolezza viewpoint at 2,921m (9,583ft). At Morteratsch, there are clear views of the Morteratsch Glacier and Piz Bernina, the highest summit in the eastern Alps at 4,093m (13,428ft). Climbing above the tree line, the train reaches its highest station, Ospizio Bernina beside Lago Blanco. If you're planning a hike, this is an excellent place to alight.

The dramatic descent begins via fir-covered slopes to the Alp Grüm viewpoint, set between Lago Palü and the glaciated peak of Piz Palü. Next, the train hairpins down to the pretty village of Poschiavo and along the valley of the same name. After the spiralling Brusio Viaduct, the Express crosses the Italian border at Campocologno and finally rolls into Tirano.

IN THE KNOW

The signature drink on the Bernina Express is a Röteli – Swiss cherry liqueur – called *Pfiff*, meaning whistle. Legend has it that as the train approached the station inn at Alp Grüm, the driver would sound the whistle, with one blast for each drink his crew wished to order.

ALSO TRY

Glacier Express, Graubünden to Valais: Switzerland's other great scenic train takes eight hours to trundle between St Moritz and Zermatt. Opt for Excellence Class and you'll receive top notch personal service.

The Top of Europe, central Switzerland: The train journey from Lucerne to Jungfraujoch, which at 3,454m (11,332ft) is the highest station in continental Europe, is another Swiss mountain classic. Travelling via Meiringen (birthplace of the meringue) and Interlaken Ost, it takes around five hours.

Arrive in style in Arlberg, the heart of the Austrian highlands

WEST AUSTRIA

Sit back, relax and hum the 'The Sound of Music' to yourself as your train glides through the mountain landscapes

Dashing over the beautiful Austrian Alps between Bludenz and Innsbruck, the 136-km (85-mile) Arlbergbahn (or Arlberg Line) is one of Europe's highest standard-gauge railways. When travelling through northern Switzerland, Liechtenstein and Austria by train, this section of the journey is a hands-down highlight. Managed by Austrian Federal Railways (OBB), both Austrian and international trains rush along these tracks. You don't need a special ticket for the experience, but can travel on all manner of trains including high-speed services linking Zürich and Vienna or the Venice Simplon-Orient-Express.

The Arlbergbahn's signature feature is the 10.6-km (6.6-mile) Arlberg Tunnel, which cuts through the Arlberg massif. It's 1,310m (4,300ft) above sea level at its highest point, with a steep climb from Langen am Arlberg followed by a gradual descent to St Anton.

The Arlberg Line was first conceived in the 1840s. As they strived to connect England to Egypt by rail, Victorian engineers considered this railway across the Arlberg Pass to be a key piece in the puzzle.

The grand opening didn't take place until the 1880s. The region presented the Arlberg's creators with countless technical problems, beset as it is by avalanches, mudslides, rockfalls and floods. In total, 92 lives were lost during the four-year construction period.

THE LOWDOWN

Best time of year: Any time. Spring is good for hikes though lush pastures dotted with wildflowers, while winter offers snowy landscapes and Alpine sports.

Plan your trip: The Zürich to Vienna railway is eight hours long. Its most scenic section, the Arlberg Line from Bludenz to Innsbruck, takes around 1hr 40min.

Getting there: Zürich and Vienna are around 940km (580 miles) and 1,470km (910 miles) respectively by land and sea from London. Eurostar trains from London St Pancras to Paris Nord connect with trains from Gare de Lyon to Zürich, taking around 7hr 30min. Returning from Vienna via Brussels or Amsterdam takes around 13hr 30min (or 17hr 30min by sleeper and Eurostar). There are buses to Zürich via Paris, and from Vienna via Prague.

Load your wheels onto the train, and go: the Arlberg Line takes you into the heart of a mountainous region that's perfect for exploring by electric mountain bike.

ACTIVE AUSTRIA

While Austria is superb for winter sports, there are plenty of reasons to visit once the snow has melted, too. If you like your maps crammed with contour lines, you'll love hiking routes such as the Tyrolean section of the E4 long-distance path, one of the best in Europe.

RIDE THE ALPEN EXPRESS

From late December to mid March, a weekly sleeper train links Amsterdam and Cologne to Austria's ski resorts including Bludenz, St Anton and Innsbruck. Leave London St Pancras by Eurostar on Friday, change onto the Alpen Express in the evening, and you'll be in the mountains on Saturday morning, ready to travel the Arlberg Line or hit the slopes.

Ski poles at the ready

If you're outdoorsy, winter is the obvious time to visit: Arlberg is arguably the birthplace of modern Alpine skiing, and St Anton am Arlberg is home to Austria's largest ski school. There are aspects of downhill skiing that are far from eco-friendly – the sport can disrupt mountain habitats, and infrastructure such as lifts and snow cannons consume large quantities of fossil fuel – but Arlberg's resorts have a better record than most. Schröcken, north of St Anton, is high enough to rely entirely on natural rather than generated snow. St Anton is self-sufficient in hydro power, and its meadows are monitored for environmental damage.

The Flexenbahn, a loop of lifts and cable cars, connects St Anton with Lech, Zürs, St Christoph and Stuben, making this one of the largest ski and snowboard regions in the world. The slopes here are superb, particularly for intermediate skiers. At other times of year, the line offers access to a network of hiking trails.

Chocolate, stamps and Sachertorte

Start your railway journey across northern Switzerland and Austria in Zürich, and you can wander the cobbled Old Town, delve into the city's art museums, relax at lakeside lidos and indulge yourself at the famous Lindt & Sprüngli chocolate factory.

To explore Vaduz, capital of Liechtenstein, Europe's fourth-smallest country, alight at the Swiss border town of Buchs and take a bus. Once you've admired the neo-Gothic cathedral, the strikingly modernist art museum and (if stamps are your bag) the Postal Museum, rejoin the railway at Buchs and continue via Feldkirch to Bludenz.

East of Bludenz, you're on the Arlberg proper, coasting through Alpine pastures and mountainscapes. Between St Anton am Arlberg and Landeck, you'll cross the Trisannabrücke, a magnificent viaduct whose iron arch soars 87m (285ft) above the River Trisanna. The Arlberg Line officially ends in Innsbruck, linking seamlessly with the next section of the main line to Vienna. Innsbruck's Old Town, immediately west of the station, has a colourful, historic character, with lederhosen and dirndls for sale. The city's most modern touch is the Hungerburgbahn funicular designed by Zaha Hadid, which climbs the Nordkette range for superb, panoramic views.

ALSO TRY

Semmeringbahn, West Austria: Completed in 1854, the World Heritage Semmering Line from Gloggnitz to Mürzzuschlag proved it was feasible to build a standard-gauge mountain railway, paving the way for the Arlberg. Only 41km (25 miles) long, it contains 14 tunnels and 16 viaducts.

Brennerbahn, Austria to Italy: The scenic Brenner Line, Austria's highest standard-gauge mountain railway, connects Innsbruck and Verona via the Wipp Valley, Brenner Pass and Adige Valley.

The winter sports resort of Lech Zürs am Arlberg generates 80 per cent of its energy from local biomass power plants and takes steps to protect its wildflower meadows from seasonal damage.

Explore the Flåm Valley by train

SOUTHWEST NORWAY

Glorious in any season, the Flåmsbana is one of the most scenic railways in the world

A railway line with a 5.5 per cent gradient? That may not sound like much. But for a standard-gauge railway – as opposed to a narrow-gauge mountain line, rack railway or funicular – it's steep. The steepest in Europe, in fact. You'd half expect to find such a railway in a white-knuckle-ride theme park, but the Flåmsbana (Flåm Railway) is in fact a regular, state-owned passenger route.

You don't take this 20.2-km (12.5-mile) electric-powered journey just for the thrill of the gradient, though – you come for the mountain scenery. It's spectacular. Branching off the Oslo to Bergen Bergensbanen (Bergen Line) at the roadless, one-reindeer town of Myrdal, 867m (2,845ft) above sea level, the Flåmsbana rushes down through the wild and beautiful Flåm Valley to the village of Flåm, at an altitude of 27m (89ft) on the Aurlandsfjord. On the way, it passes forests, farmhouses, waterfalls and the gushing River Flåmselvi. At the Kjosfossen waterfall, which is only accessible from the railway, the train stops for everyone to get out and take photos, while folk music plays and a young Norwegian dressed as a Huldra, a siren-like Norse forest spirit, performs a dance close to the water's edge.

Fine landscapes and dancing nymphs aside, the railway itself is impressive, from its 1980s El17 electric locomotives and smart carriages with panoramic windows to its twenty tunnels that spiral in and out of the mountains. All but the two longest tunnels – 889m (2,917ft) and 1,342m (4,401ft) respectively – were built by hand using drills and dynamite in the 1920s and 1930s.

THE LOWDOWN

Best time of year: Any time. High season is May to September when there are ten departures in each direction per day. November to March are the quietest months, with four return trips per day.

Plan your trip: A ride on the Flåmsbana takes 45–75 minutes each way from Myrdal, which is five hours west of Oslo and two hours east of Bergen by Bergensbanen train. Allow one or two weeks for a complete 'Norway in a Nutshell' tour.

Getting there: Oslo, a good base for this trip, is around 1,730km (1,080 miles) by road and sea from London. By train, travel via Copenhagen (see page 46). From here, either continue to Oslo by train via Sweden (9hr), or take the ferry north across the Kattegat and Skagerrak (19hr). From London, allow at least 27hr (or 38hr if you opt for the ferry).

Undredal, a fjord-country village known for its goat's cheese, is around 30 minutes from Flåm by ferry, following the banks of Aurlandsfjord.

In the village of Flåm, the railway office, visitor centre and neighbouring Flåm Railway Museum are open dailys.

IN THE KNOW

For the best views as you ride the Flåmsbana from Myrdal down to Flåm, try to find a seat by a left-hand window. Heading uphill, sit on the right.

The Flåm Valley is excellent for eco-friendly outdoor activities. You can go snow-shoeing or boating, pelt along Scandinavia's longest zipline or hire a small electric car and explore – perhaps zooming up to Stegastein for views over the Aurlandsfjord. The Fretheim Hotel in Flåm serves local specialities such as cured reindeer with green strawberries, cloudberry soup, *gjetost* (brown-coloured goat's cheese) and *aquavit* (the potato-based spirit that Scandinavians love to drink at Christmas). Hiking back up from Flåm to Myrdal, rather than returning by train, is possible: allow a full day.

Norway in a Nutshell and the Rallarvegen

The Flåmsbana fits neatly into a train trip between the attractive, green cities of Oslo and Bergen, perhaps following the classic 'Norway in a Nutshell' route, which visits the region's best landscapes by train, bus and battery-powered boat. Don't miss the two-hour eco-friendly cruise to the Viking village of Gudvangen through World Heritage-listed Aurlandsfjord and Nærøyfjord, a highlight of the experience.

The 371-km (231-mile) Oslo to Bergen Bergensbanen, northern Europe's highest railway, is itself superbly scenic. As well as stopping at Myrdal, it serves ski resorts Gol, Geilo and Voss. It also stops at Haugastøl, head of the classic Rallarvegen (Navvies' Road) cycling route, which follows the railway via Finse to Myrdal then hairpins downhill alongside the Flåmsbana to Flåm.

THE RIGHT TO ROAM

Norwegians have a philosophy called *allemannsretten* meaning that you are free to roam nearly anywhere as long as you respect the natural environment by being considerate and thoughtful and leaving the landscape as you find it.

ALSO TRY

Dovre Line, Norway: The seven-hour train journey from Oslo to Trondheim immerses you in glorious national park scenery. On the way, you'll pass through the Gudbrandsdalen valley and across the mighty Dovrefjell Mountains.

Nordland Line, Norway: This famous route runs between Trondheim and Bodø, crossing the Arctic Circle. At 729km (453 miles), it's Norway's longest railway.

Santa Claus Express, Finland: Experience the magic of midwinter as you journey by night train from Helsinki to Rovaniemi in Lapland.

Travel like a tsar from Sofia to the Black Sea

BULGARIA

A supremely nostalgic railway journey, travelling through the Balkan Peninsula by steam train

Imagine exploring southeastern Europe not by budget airline, but by royal train. On board Bulgaria's Korona Express, you can travel in style across the Thracian Plain and the rugged Rhodope Mountains in 1930s carriages that belonged to Tsar Boris III, drawn by magnificently restored steam locomotives dating back to the 1940s. Organized railway holidays spin out the journey nicely, combining days on the train, or on excursions, with nights in hotels.

Train buffs adore the Korona Express: its itineraries have plenty of aficionado-friendly touches built in. Expect a good variety of stops at depots, junctions and stations. You'll even get the chance to alight, temporarily, for 'run past' photo opportunities, watching the train wend its way through particularly dramatic locations.

But if all you want to do is enjoy Bulgaria's varied countryside, you'll be perfectly at home: there's much for window-gazers to enjoy. As the train pulls out of Sofia Central Station with a cheery whistle, settle down in a walnut-panelled compartment in preparation for the sightseeing to come. Soon, the suburbs give way to rural areas, with onlookers waving and snapping photos as you chug through hillside villages.

Your first destination is Gorna Oryahovitsa, the biggest railway centre and steam locomotive depot in northern Bulgaria. It's a jumping-off point for Veliko Tarnovo, the nation's medieval capital, where houses teeter on cliffs and a 12th-century hilltop fortress, Tsarevets (*pictured, left*), looms over the River Yantra.

Back on the Korona Express, the scenery changes as you travel via Dabovo and Stara Zagora through valleys and lowlands up into the Balkan Mountains, then down to Burgas, from where the beach resort of Nesebar, Pearl of the Black Sea, is within easy reach.

THE LOWDOWN

Best time of year: Any time. Peak holiday season is June to August.

Plan your trip: The train journey between Sofia and Burgas takes around seven to eight hours each way. Railway holiday operators charter the Korona Express for return trips lasting around ten days, with stops and excursions en route.

Getting there: Sofia is around 2,480km (1,540 miles) by road and sea from London via Germany and the Balkans. Eurostar trains from London St Pancras connect with trains via Brussels and Frankfurt to Linz in Austria, for a bus to Sofia. Allow at least 30 hours, including 16hr 30min on the bus. Alternatively, if you're up for a longer journey (over 40hr) with multiple changes, you could train-hop all the way from London.

Back to Sofia via the Rhodope Mountains

Heading west once more, the train visits the spa town
of Velingrad in the Rhodope Mountains, where you
can take a side trip to Septemvri to ride a powerful
little 1940s Polish narrow-gauge steam train up the
gorge of the River Cepina.

The next leg is a real treat: a trip on the narrow-gauge
line that climbs from Velingrad to Avramovo, the plume
from the engine billowing as it tackles the steep ascents
and superbly engineered spiral tunnels. At 1,267m
(4,157ft), Avramovo is the highest station in Bulgaria,
lying close to the summit of the Rhodope Mountains.

From here, it's on to the ski resort of Bansko and finally
to Dobrinishte. This is a superbly scenic day on the
tracks, rattling through forested gorges, ravines and
bucolic highland meadows. Your final excursion is to
the UNESCO World Heritage site of Rila Monastery;
then the Express carries you back to Sofia.

ALSO TRY

The Vitosha Express:
Like the Korona
Express, this 1970s
train can be chartered
for tours of Bulgaria's
railways. It was
designed for
communist head of
state Todor Zhivkov,
and retains many of
its original features.

Belgrade to Bar:
Soak up the views on
another of the Balkan
Peninsula's most scenic
train journeys. This
12-hour trip between
Serbia and Montenegro
travels through the
foothills of the Dinaric
Alps to the Adriatic via
canyons and ravines.

*Fishing village on the Vacha
Reservoir in the Rhodope
Mountains, where Bulgaria
generates a significant
proportion of its hydroelectricity.*

ADVENTURES IN THE SLOW LANE

Sometimes, life feels far too much like a race. So how about a holiday that invites you to pause, recharge, and maybe even turn back time? A trip down memory lane to a sandy bay, perhaps. A saunter through medieval villages. Or a jaunt in a gypsy-style caravan. Forget soaring like an eagle or dashing around like a hare. Be more tortoise. It's a wonderful way to explore.

Backpack along the Wales Coast Path

WALES, UK

Wales is the only country that invites walkers to explore its entire coastline, visiting castles, beaches and islands dotted with puffins

In a world where people jostle for a foothold on the property ladder and a sea view is a precious commodity, any plan to build a coastal footpath for everyone to enjoy could be considered a victory for fairness, health and nature.

Well done Wales: it's the first country to take this notion and run with it, creating a public footpath that follows every inch of its beautiful coastline, staying close to the shore for much of the way. The Wales Coast Path was many years in the making, with the last gates, stiles and 'dragon-shell' signposts put in place in 2012. Totalling a little over 1,400km (870 miles), this ambitious path is perhaps longer than you'd expect, but remember: Wales has a crinkled, craggy coastline, full of rugged headlands, breezy bays, hidden inlets and secret coves.

Between the official endpoints – Chepstow in the south and the River Dee near Chester in the north – the countryside is walker-friendly and superbly varied, much of it lying within national parks, nature reserves or areas of outstanding natural beauty. While the more rugged sections are uneven, with steep climbs to negotiate, many stretches are suitable for cyclists, pushchairs, wheelchairs and horse riders, and these are indicated on the Wales Coast Path website and local maps. For those who might relish the challenge of circumnavigating an entire country, the endpoints are also linked by an inland national trail that follows Offa's Dyke along the land border between Wales and England, closing the loop.

Trace the shape of the nation

Walk the entire Wales Coast Path from north to south, and you'll start on an upbeat note with the seaside towns of north Wales, followed by Conwy's magnificent castle and Bangor, with the option of crossing the Menai Strait to circumnavigate Anglesey. Next come Caernarfon, the Llŷn Peninsula and Cardigan Bay, past Harlech, Aberystwyth and Cardigan. The magnificent Pembrokeshire Coast is a highlight, as is the Gower Peninsula. Finally it's on to lively, likeable Swansea and Cardiff, before the big finish in Chepstow.

The view from the Wales Coast Path to Gower's Three Cliffs Bay – a sandy beach that's only accessible on foot – is considered one of Britain's best.

COAST PATH HALL OF FAME

Soon after the Wales Coast Path opened in 2012, adventurers began setting themselves challenges. Runner Arry Beresford-Webb was the first to run the entire path, completing it at the rate of one marathon a day, and Hannah Engelkamp walked it with a seaside donkey.

If you'd prefer to focus on one section at a time, staying overnight at hostels, B&Bs or campsites, you'll find plenty of options with start and end points close to towns, railway stations or bus stops. Offshore, protected islands such as Skomer, Skokholm and Bardsey, accessible only by small, open-deck passenger ferry, are summer breeding sites for seabirds galore. With colourful, characterful puffins, predatory black-backed gulls and migratory shearwaters, razorbills and fulmars wheeling overhead, perching on ledges or taking refuge in rabbit burrows, these islands offer some of the best bird-watching experiences in the UK.

Opposite: the tip of the Llŷn Peninsula points towards Bardsey, a genuinely peaceful island with a bird observatory and a few rustic cottages, available to rent from March to October.

Three regions to explore

Llŷn Peninsula

Reaching out towards Ireland from Snowdonia, the beautiful, remote Llŷn Peninsula is fringed by farmland, cliffs, fishing villages, sailing marinas and a 161-km (100-mile) path. A minibus service, the Llŷn Coastal Bus, helps make this underrated rural region accessible to walkers. Operating between late March and October, it connects up the towns and can be pre-booked or hailed, space permitting.

Pembrokeshire and Carmarthenshire

A national park since 1952, the 290-km (180-mile) Pembrokeshire coast is a jewel in the Welsh crown. In spring, on cliffs tufted with harebells and thrift, you'll hear lambs bleating and see choughs tumbling on the breeze. Sometimes, the path disappears into woods draped with wild honeysuckle; often, it leads to perfect swimming beaches with golden sand. Hail-and-ride public minibuses (the Pembrokeshire Coastal Bus) zip along the lanes in summer.

Carmarthenshire Bay and the Gower Peninsula

Come here for huge, wild beaches. The sands at Rhossili and Oxwich on the 70-km (43-mile) Gower Coast Path are regularly voted among Britain's best. It's also a region with poetic associations: the great Welsh writer Dylan Thomas lived for two decades, on and off, in Laugharne, and the village is now a place of literary pilgrimage.

At Rhossili on the Gower Peninsula, a steep descent leads down to broad, breezy sands lapped by the Bristol Channel, drawing surfers and bird-watchers as well as walkers.

ALSO TRY

Great Ocean Walk, Australia: Instead of driving Victoria's celebrated Great Ocean Road, you could walk part of the same coast. This path spans 100km (62 miles) between Apollo Bay and the Twelve Apostles.

Whale Trail and Otter Trail, South Africa: These famous coastal hiking routes take five or six days each, with overnight stops in cottages. The Whale Trail explores the De Hoop Nature Reserve; the more adventurous Otter Trail connects Storms River Mouth and Nature's Valley. Permits must be booked in advance.

Put your best foot forward on the West Highland Way

WEST SCOTLAND, UK

A strong contender for Britain's finest long-distance walk, this much-loved trail leads through spectacular mountain scenery

Stretching 154km (96 miles) from Milngavie on the northern edge of Greater Glasgow to Fort William in the Scottish Highlands, the West Highland Way, one of Scotland's designated Great Trails, is a classic. Set in gloriously wild landscapes, it offers front-row views of blockbuster natural attractions such as Loch Lomond, Rannoch Moor, Glen Coe and Ben Nevis.

Even though it's demanding in places, the West Highland Way is Scotland's most popular long-distance hike, with around 36,000 people walking the entire route each year. That may sound like a worryingly large number, but don't be put off – if you avoid the summer holidays, you're likely to see more shaggy highland cattle and wild goats than humans on your journey.

Clearly signposted with thistle-in-a-hexagon markers, making navigation easy, the route follows old drovers' paths, military roads, abandoned railways and country tracks. It's linear, and the usual way to tackle it is from south to north, to keep the prevailing southwesterlies at your back and give you time to get into your stride. At the northern end, the terrain is hillier and the gaps between overnight stops longer. Underfoot, the surface is mostly gravel or stony earth; it's rough in places, but equipment-wise, you don't need anything more specialsized than decent walking boots and perhaps a pair of hiking poles. A backpack with some lightweight all-weather layers, blister treatment, food and water is essential, too, since much of the path is remote.

THE LOWDOWN

Best time of year: April to October. For mild weather and relatively few midges, June and September are ideal.

Plan your trip: Allow seven or eight days to complete the entire trail from Milngavie (pronounced *mull-guy*) near Glasgow to Fort William on Loch Linnhe: longer if you'd like to extend certain overnight stops or climb Ben Nevis.

Getting there: Milngavie and Fort William are around 670km (410 miles) and 820km (510 miles) respectively by road from London. By daytime trains, Milngavie is around 5hr 25min from London Euston, 4hr 30min from Manchester Piccadilly or 25min from Glasgow Central. Fort William is 9hr from London, 7hr 20min from Manchester or 3hr 20min from Glasgow. Both Glasgow and Fort William are also on the Caledonian Sleeper route from London, and are well served by buses.

Above Kinlochleven, the West Highland Way offers dramatic views of Loch Leven, one of the most scenic sea lochs in Scotland.

RUCKSACK OR DAYPACK?

Hardened backpackers can walk the West Highland Trail independently, wild camping or hostel-hopping. If you'd prefer to travel light, ask a walking-tour company to arrange daily accommodation, meals, taxis and luggage transfers for you, enabling you to walk with just a daypack. Guided group walks are also available. There are plenty of suggestions on the official website.

KNOW THE RULES

Wild camping is legally permitted in Scotland, but only if practised responsibly: you should leave no trace. In national parks, an exception applies between 1 March to 30 September. At this time, you may only camp in designated zones, which should be booked in advance. This affects the section of the West Highland Trail that passes Loch Lomond.

Walking the trail

Once you've left Glasgow behind, the walk starts gently, leading through tussocky pastures and woodlands. Next, it winds along Loch Lomond's long, lovely eastern bank, with Inversnaid Falls plunging through the trees. North of the loch, there's a dramatic change in the landscape as you enter the Highlands proper. Here, the path is edged with heather, rowan and gorse and the mountains are more substantial.

The scenery gets even more spectacular once you're passing Loch Tulla and crossing Rannoch Moor. You'll need to summon your best leg power to climb the Devil's Staircase, built by soldiers in the 18th century. Zigzagging up out of Glen Coe, it rewards you with glorious views of moorland and mountains.

Your final day brings yet more wonderful vistas, as you gaze across Loch Leven and, if the weather is kind, up at mighty Ben Nevis. At 1,345m (4,413ft), it's the highest mountain in the British Isles. If it brings out the mountaineer in you, now's your chance: the path to the summit begins 3km (2 miles) southeast of Gordon Square in Fort William, where the West Highland Trail ends.

For a relaxing diversion from the route, wander the trails around Glencoe Lochan, a lake surrounded by ornamental forest, just north of the village of Glencoe.

ALSO TRY

Great Glen Way, Scotland: This 125-km (78-mile) great trail across the Scottish Highlands from Fort William to Inverness is ideal for those who, on completing the West Highland Way, still have energy to spare.

Ulster Way, Northern Ireland: This 1,000-km (621-mile) national trail is an epic circular walk. It connects several areas of outstanding natural beauty, including the Sperrin Mountains and Causeway Coast.

Cleveland Way, England: Manageable in nine days, this 175-km (109-mile) national trail loops across the North York Moors and along the Yorkshire coast, through some of northern England's classic landscapes.

Follow the Travellers' trails of County Wicklow

EAST IRELAND

THE LOWDOWN

Best time of year:
May to September.
High season is July
and August.

Plan your trip: Self-
catering horse-drawn-
caravan holidays are
typically three nights
long. Donkey walking
tours, staying at village
B&Bs, are up to seven
nights long.

Getting there: County
Wicklow is around
650km (400 miles)
by road and sea from
London via Wales, and
60km (37 miles) by road
from Dublin. Travelling
to Dublin by train from
London Euston and
ferry from Holyhead
takes around eight
hours. For other routes,
see page 54. Buses from
Dublin to Wicklow take
a little over an hour.

Take to the open road in a brightly painted caravan, drawn by an Irish cob horse

When Kenneth Grahame's Toad of Toad Hall from *The Wind in the Willows* treated himself to 'a gypsy caravan, shining with newness, painted a canary-yellow picked out with green, and red wheels', he was flushed with the wanderlust that many of us feel from time to time. 'Here to-day, up and off to somewhere else to-morrow!' he cried. 'Travel, change, interest, excitement! The whole world before you, and a horizon that's always changing!'

Visitors to County Wicklow who might be pondering a back-to-nature stay in a teepee, Mongolian yurt or shepherd's hut can choose a mobile glamping escape instead, if they wish: living the romantic rural dream, Toad-style, in their very own gypsy caravan.

Like Toad's friend Mole, when he first saw the canary-coloured cart, you may well fall in love with your wagon at first sight. As in the story, you'll find it compact and comfortable, with bunks, bookshelves, pots, pans, kettle and a cooking stove (though not a birdcage with a bird in it). You'll have all you need, in a contraption that's light enough to pull.

Importantly, it comes with a horse – an Irish cob, known for being strong and docile – and you can choose to travel with or without a driver (who would double as a knowledgeable local guide). It's not necessary to have any prior horse-handling experience; you'll be offered a horse that suits your ability and will spend your first day learning how to feed, groom, catch, harness, yoke, drive and shower it.

To extend your stay in County Wicklow, allow some time in the Wicklow Mountains National Park, for serene waterside walks.

GARDEN OF IRELAND

County Wicklow is known as The Garden of Ireland: high praise, in a country as lush as the Emerald Isle. Botanists have recorded more than 800 plant species here, including rarities such as parsley ferns and bog orchids. In spring and summer, look out for wildflowers such as blackthorn, herb bennet, cowslips and buttercups.

The heath, the common, the hedgerows...

Most of Ireland's traditional travellers are practising Catholics with strong beliefs in the power of prayer. They're bilingual in English and Shelta, Cant or Gamin, dialects, which they prefer not to share with outsiders, and their work – as horse traders, farm labourers and entertainers – fits around their itinerant lifestyle.

A gypsy caravan holiday doesn't attempt to explore this lifestyle in any depth, but simply borrows a few of its freedoms. Your caravan will probably lack a bathroom, so you'll use it for short, easy-going day trips. There are walks to enjoy and traditional pubs to visit; on occasion, you might choose to give your horse a rest and set off on a walk with one of the owner's donkeys instead. Come evening, you'll camp on a nearby farm or simply return to the owner's fields, building a campfire for a barbecue under the stars.

As your jaunty little caravan wends its way along quiet lanes surrounded by rolling green countryside – a patchwork of fields and leafy valleys, crossed by clear rivers and pretty stone bridges – all you'll hear for much of the time are bird calls and the clip-clop of hooves. 'There's real life for you,' as Toad so rightly said.

The leafy slopes, beer-brown waters and picturesque ruined abbey of the Glendalough Valley in the Wicklow Mountains National Park are within easy reach of Dublin.

ALSO TRY

The Wicklow Way, Ireland: Eire's oldest long-distance walk is a 130-km (80-mile), seven-stage route from Dublin to Clonegall. It crosses the Wicklow Mountains via Powerscroft Waterfall (Ireland's highest), Lough Tay (nicknamed the Guinness Lake for its dark water) and the gorgeous valleys of Glenmalure and Glendalough.

Wagon tour, Vale of Eden, England: Take a horse-drawn wagon along country lanes between the North Pennines and the Lake District.

Glamping, Haut-Beaujolais, France: Stay in an elaborately decorated *roulotte*, or French showman's caravan, parked in a grassy garden.

Bird-watch by bike, from Denmark to Belgium

DENMARK, GERMANY AND LOW COUNTRIES

Pedal along the North Sea coast, one of Europe's richest habitats for migratory water birds

Denmark, Germany, the Netherlands and Belgium are among Europe's most cycle-friendly destinations, so why not link them together on an easy-going long-distance adventure? Of all the routes you could take, one of the most rewarding is the 2,300-km (1,430-mile) central section of the EuroVelo 12 (EV12), or North Sea Cycle Route, from Grenaa in Denmark to Bruges in Belgium. It follows the coast of the Kattegat, Skagerrak, German Bight and North Sea. For keen bird-watchers, it's a delight: the only hitch is that you'll be constantly stopping to pull out your binoculars.

EuroVelo is a network of sixteen long-distance European cycle routes developed by the European Cyclists' Federation (ECF). Created with both locals and tourists in mind, the routes range in length from 2,050–11,000km (1,274–6,835 miles). Each route connects several countries and runs reasonably close to railway stations, allowing you to tackle short sections if you wish.

EV12 is one of the most ambitious EuroVelo routes, hugging the North Sea coast all the way from Bergen in Norway to Lerwick in Scotland's Shetland Islands, with brief forays to Hamburg and through parts of inland England and Scotland. When it opened in 2001, it was still possible to a catch a passenger ferry between Lerwick and Bergen, closing the loop and earning EV12 a Guinness World Record for being the world's longest cycle route.

THE LOWDOWN

Best time of year: April to October.

Plan your trip: You could cycle a short section of the route, such as Delfzijl to Amsterdam, in around five days. To complete the entire 2,300km (1,430 miles) from Grenaa to Bruges, allow thirty days or more.

Getting there: Grenaa and Bruges are around 1,320km (820 miles) and 280km (175 miles) respectively by land and sea from London via Kent. Trains from London St Pancras via Hamburg and Århus to Grenaa take around 24hrs, with an overnight stop. Ferries from Dover, Harwich, Hull or Newcastle to France, Belgium or the Netherlands link to Denmark by road and rail. Bruges is 3hr 30min from London by Eurostar train. There are stations near many intermediate points. Rules about bikes on trains vary, so check in advance.

Waterfowl in flight over the Frisian island of Langeoog, which lies within Germany's Lower Saxon Wadden Sea National Park.

EV12 FOR BEGINNERS

The easiest part of the route for beginners and young families is the Dutch section. Here, the paths are particularly wide, smooth and well signposted. Many are virtually traffic free.

IN THE KNOW

Long-distance cyclists either carry camping gear, or hop between places to stay that are geared up to welcoming them. Aktiv Danmark's handy Bed + Bike certification identifies cyclist-friendly hotels, B&Bs and hostels near popular routes; many offer storage, repair kits and other useful extras. The German and Dutch counterparts are Bett und Bike and Fietsers Welkom. The Netherlands has a flat-rate homestay network for cyclists, Vrienden op de Fiets.

Coastal discoveries

Today, with no ferries between Norway and Scotland, the North Sea Cycle Route is U-shaped rather than circular, but it's no less interesting for that. For much of the Danish, German, Dutch and Belgian sections, you'll be riding past the biodiversity-rich dunes, tidal flats and wetlands of the Frisian Islands and Wadden Sea coast, a UNESCO World Heritage listed protected area and transfrontier national park. This, the largest unbroken system of intertidal sand and mud flats in the world, is a crucial breeding and feeding ground for resident and migratory birds, including white-tailed eagles, herons and thousand-strong flocks of waders, ducks, geese, gulls and terns.

Other natural highlights include Denmark's Hvide Sande, Tipperne peninsula and Holmsland Klit (*pictured, left*). In this remarkable area, mighty dunes separate Ringkøbing Fjord, a shallow lagoon, from the North Sea, creating one of northern Europe's largest bird habitats.

Between The Hague and Bruges, you ride through another precious wetland region, Oosterschelde National Park, the largest protected area in the Netherlands. Again, this is a region where saltwater and freshwater merge: the grass-tufted dunes and flats alter with each ebb and flow of the tides. Rich in natural food, the park attracts dozens of water birds, fish and crustaceans, along with seals and harbour porpoises.

When it's time to take a break from bicycling and birds, you'll find much to do in the region, from visiting Danish whisky distilleries to exploring the cosmopolitan culture and historic architecture of Hamburg, Amsterdam, The Hague and Bruges.

ALSO TRY

EuroVelo 6, the Atlantic to Black Sea route: One of Europe's most popular cycling routes, it runs from Nantes, France, to Constanta, Romania, and includes the Danube Cycleway, following the River Danube through southern Germany to its delta in the Black Sea.

EuroVelo 13, the Iron Curtain trail: This is the longest EuroVelo route at 9,950km (6,182 miles). It covers a thrilling mix of landscapes, passing through rural Scandinavia and along the Baltic coast, then tracing the old Iron Curtain for a living lesson in European history.

Mess about in an electric canal boat in Alsace

NORTHEAST FRANCE

Enjoy the sounds of rippling water and birdsong as you pilot a motor cruiser that's as clean and quiet as a rowing boat

It's high time the world of canal boating had an eco-conscious shake-up. Why don't all cruisers and narrowboats run on something cleaner and greener than diesel, gas and coal? Often lined with mature trees, hedgerows, farmland, hillsides or wetlands, Europe's canals are precious wildlife corridors that surely deserve better than to be used as motorways for noisy, polluting fuel guzzlers.

While electric boating is by no means new – silent, battery-powered vessels have been around since the 19th century – it's taking time for modern boatbuilders and cruising enthusiasts to make the switch to renewable energy. Those that have succeeded have revolutionized cruising, transforming it into slow travel at its best. Leisurely and convivial, eco-cruising is a perfect partner to wildlife-watching, inn-hopping, towpath walks, bike rides and good old-fashioned relaxation, sitting on the prow and letting the countryside glide by.

One of the boat hire outfits leading the way has bases at Saverne and Harskirchen on the Marne-Rhine Canal in the northern Vosges Mountains. Here, you can rent a custom-built Sixto Green electric cruiser that sleeps six, for a zero-emissions trip. Free of engine noise and vibration, a Sixto Green is a perfect platform for bird-watching and photography and, with no generator, evenings are peaceful, too. A two-hour charge of the lithium-ion batteries can last up to eight hours, more than enough for a typical day's itinerary, and the Marne-Rhine has towpath charging points every 11km (7 miles), so you should never run out of juice.

Going up in the world

The Plan Incliné de Saint-Louis Arzviller on the Marne-Rhine Canal, west of Saverne, is an engineering masterpiece. This counterweighted boat elevator raises vessels 45m (150ft) in altitude. Installed in 1969, it replaced a sequence (or ladder) of 17 locks that used to take a whole day to negotiate. The Plan Incliné takes just twenty minutes.

A leisurely 100 kilometres (60 miles) west of Saverne, the Marne-Rhine Canal passes through the historic town of Nancy.

THE LOWDOWN

Best time of year: May–June and September. High season is July and August. Boat-hire companies offer very limited services between November and March.

Plan your trip: Allow three to four days for a one-way trip on the Marne-Rhine Canal between Saverne and Harskirchen, which are around 42km (26 miles) apart, or three weeks for a journey along the canals of Alsace-Lorraine, Germany and Luxembourg.

Getting there: Saverne is near Strasbourg and around 750km (470 miles) by land and sea from London via Kent. Eurostar trains from London St Pancras to Paris Nord connect with trains from Gare de l'Est via Nancy to Saverne, taking around 5hr 50min. Alternative routes with road or rail links to the region include ferries from Portsmouth, Newhaven, Harwich, Hull and Newcastle to France, Belgium and the Netherlands.

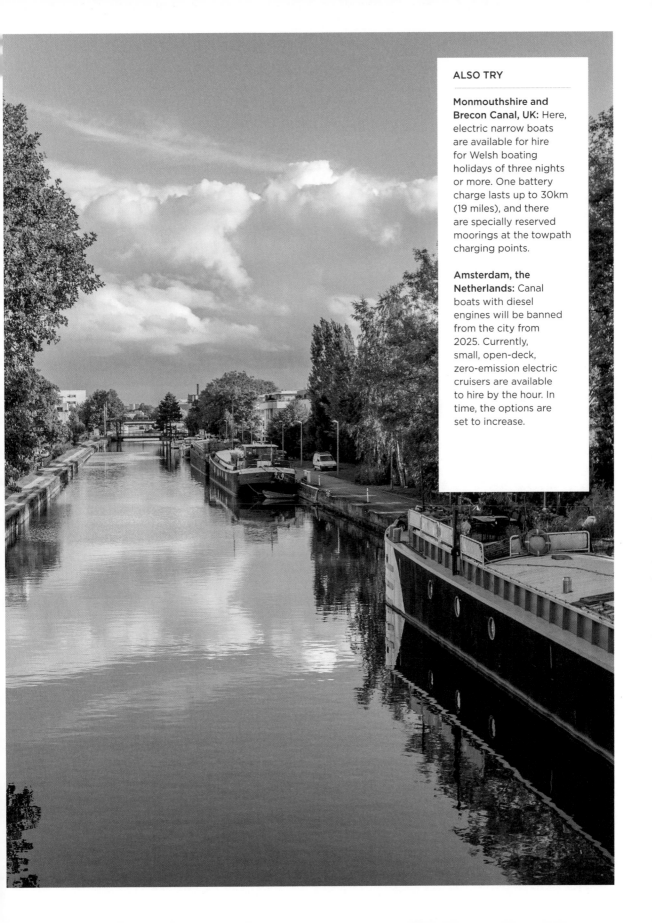

Go cross-country skiing in the French Alps

SOUTHEAST FRANCE

Grab your all-weather kit and hop on an international train: it's the coolest way to reach Europe's leading winter resorts

Some ski for the thrill of it; others ski to de-stress. If you're one of the latter, you'll know that navigating crowded airports with cumbersome gear can wreck the vibe. Fortunately, there's an overland alternative – catching a train, and hopping off within striking distance of a top French resort. It's a dream and, door-to-door, it doesn't even take much longer than flying.

If you were to start your journey in London or Kent, for example, you could crunch your travel time down to nine hours. Set off on a Friday or Saturday morning, spend the day watching the scenery fly by and you'll arrive in time for your first weekend fondue. At the time of writing, the much-loved Eurostar ski trains from London to Haute-Savoie in France are not operating, but if they resume, you'll have the option of travelling overnight – a good way to squeeze the most value out of a short break.

Travelling to a ski resort by train reduces your carbon footprint, compared to flying. But there's an elephant in the room: the impact of your stay. Some downhill ski resorts scar natural landscapes by cutting down trees and building structures that harm wildlife. Many use snow cannons, which require colossal water and energy supplies. To minimize environmental damage, choose cross-country skiing – in French, *ski de fond*. It's not perfect, but it requires relatively little infrastructure, which helps.

THE LOWDOWN

Best time of year: Late December to early April.

Plan your trip: Allow at least four days, including travel time.

Getting there: The Trois Vallées region is around 1,120km (700 miles) by land and sea from London via Kent. Eurostar trains from London St Pancras, Ebbsfleet and Ashford to Paris Nord connect with high-speed trains from Paris Lyon to Moûtiers-Salins, Aime-la-Plagne and Bourg-Saint-Maurice (from 8hr). Buses connect these stations to the resorts (30min to 2hr). Alternative routes with road or rail links to the region include ferries from Portsmouth, Newhaven, Harwich, Hull and Newcastle to France, Belgium and the Netherlands.

Compared to lower-altitude winter resorts, destinations such as Val Thorens in France have less need to top up their pistes with noisy, energy-consuming snow cannons.

GREEN SCENE

Since 1 January 2020, Eurostar, in partnership with the Woodland Trust, Reforest'Action and Trees for All, has been planting a tree for every train service it operates throughout the network. The aim is to boost European woodlands with 20,000 new trees each year.

With almost all of its pistes situated between 1,800–3,230m (5,905–10,597ft) above sea level, the Trois Vallées region only has to use artificial snow-making equipment relatively infrequently. This helps make it more eco-friendly than lower-lying resorts.

France's Trois Vallées region, the largest skiing destination in the world, is home to eight famous ski resorts including Val Thorens, Courchevel, Méribel and Les Menuires. Along with 35,000 hectares (135 square miles) of slopes for downhill skiing and snowboarding, the region has 120km (75 miles) of waymarked cross-country skiing trails, leading across frozen lakes and through wintry forests amid striking mountain scenery.

The Méribel Reforest'Action Valley is the best place to go. If you're a beginner to cross-country skiing, head to Méribel-Mottaret: here, you can get into your stride by skiing across the safe, flat surface of Lac de Tuéda, which is situated in a quiet nature reserve, surrounded by Swiss pines. When you're ready to fuel up, head for the Restaurant du Lac de Tuéda, a rustic eatery in a lakeside chalet, which serves three varieties of cheese melt – a traditional Savoie *tartiflette* made from reblochon cheese and potatoes, raclette with onion and bacon or a *Jura boîte chaude* (baked soft cheese) with *charcuterie* – along with coffee, *citron chaud* (hot lemon) and mulled wine.

With more experience, you could try the blue (intermediate) or red (advanced) trails around Méribel Altiport, which loop through the pines between Méribel and Courchevel.

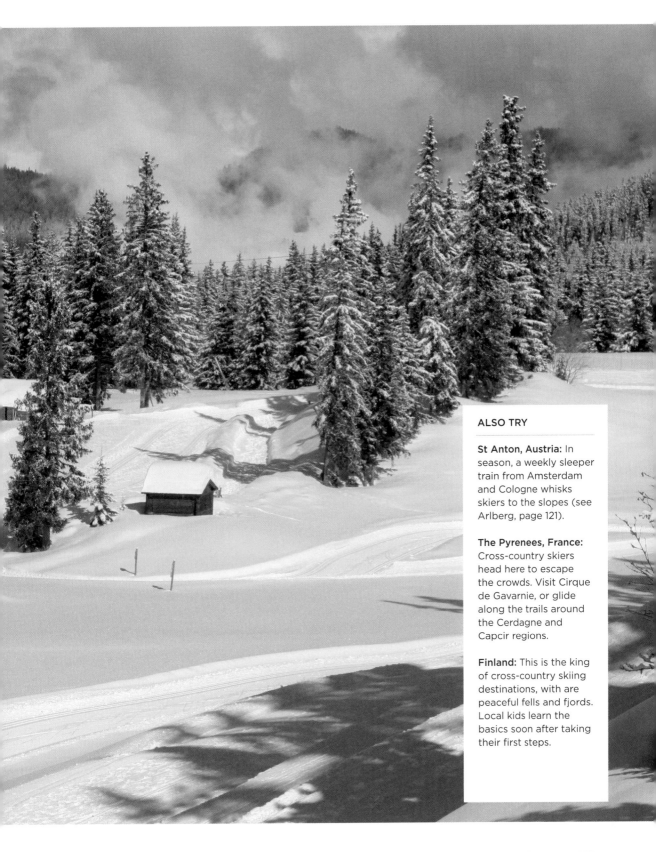

ALSO TRY

St Anton, Austria: In season, a weekly sleeper train from Amsterdam and Cologne whisks skiers to the slopes (see Arlberg, page 121).

The Pyrenees, France: Cross-country skiers head here to escape the crowds. Visit Cirque de Gavarnie, or glide along the trails around the Cerdagne and Capcir regions.

Finland: This is the king of cross-country skiing destinations, with are peaceful fells and fjords. Local kids learn the basics soon after taking their first steps.

Stride along El Camino de Santiago, the pilgrims' route

FRANCE, PORTUGAL AND NORTHERN SPAIN

Offering friendly hospitality and time to reflect, this long-distance walk is an uplifting experience, whatever your religious beliefs

In the 9th century, a hermit called Paio (Pelayo in Castilian Spanish, or Pelagius in Latin), followed mystical lights glimmering in the sky which led him to a remote oak grove. Here he discovered an old Roman burial ground containing a tomb that he believed was that of the apostle Santiago (Saint James the Great). Devoted Christians have been journeying to the spot ever since.

Historians now doubt that the remains Paio found were indeed those of Saint James, but that hasn't shaken Santiago de Compostela's lasting magnetism. At the time of the discovery, Spain was dominated by Muslims, making the find immensely significant. Before long, Christian pilgrims had started making their way to the site to pay their respects, and a chapel was built over the tomb. Construction of the Cathedral of Santiago de Compostela began in 1075, and by the 12th century, it was second only to St Peter's Basilica in Rome in significance within the Catholic Church. Expanded and embellished over the ages, it's a magnificently ornate mash-up of Romanesque, Gothic and baroque styles, with a portico stuffed with sculptures, an incense-scented nave and an elaborate gilded altar.

In the footsteps of Saint James

Whether you're walking to Santiago de Compostela as an act of faith, as a secular act of contemplation or for recreation, the cathedral is the traditional end point to the pathway, or *camino*. There is not, however, a single official route to get there, but dozens, with numerous starting points in Spain, Portugal and France. Over the years, locals living along these ancient pathways have built churches, cathedrals, hospitals, hostels and bridges to help pilgrims along their way.

If you're in it for the religious achievement, you'll need to begin your walk at least 100km (62 miles) away from Santiago de Compostela and pick up a stamped Pilgrim Passport at the local Pilgrim's office or church. Produce this at every inn or hostel you eat or stay at, and you'll receive a stamp as proof of your progress. You're required to gather at least two per day. When you arrive at the shrine, you can use this Pilgrim Passport to claim a certificate at the Pilgrim's Reception Office in Santiago de Compostela, and your name will be read out at the daily Pilgrims' Mass, held at noon.

THE LOWDOWN

Best time of year: May to June and September to October. Easter is always busy, and in the high season, July to August, the weather can be uncomfortably hot.

Plan your trip: Allow five days to walk from Sarria in northwest Spain, or five weeks for a longer route such as the Camino Francés (from Saint-Jean-Pied-de-Port in southwest France).

Getting there: Sarria is 1,870km (1,160 miles) by land and sea from London. The journey by train via Paris, Barcelona and Madrid takes around 22 hours, excluding stopovers. Saint-Jean-Pied-de-Port is 12hr 30min by train from London via Paris. Alternatively, take a ferry from Plymouth or Portsmouth to Santander or Bilbao (18–24hr), continuing on foot, or by trains and buses to Sarria (8–10hr) or St-Jean (4–6hr). Returning from Santiago de Compostela to London by train takes 21hrs.

AT THE SIGN OF THE SCALLOP

The emblem of Saint James and the Camino is the scallop shell (*concha de vieira* in Spanish, *coquille Saint-Jacques* in French). You'll spot them repeatedly on the route – brass ones embedded in pavements, ceramic ones in walls – their grooves converging on a point. Traditionally, local *hospitaleros* (hosts) who spotted a walker with a shell attached to their walking stick or backpack would offer them hospitality, a custom that continues today.

KNOW YOUR MENU

If the sight of all those scallop shells gives you an appetite for seafood, you'll love popping into local Galician eateries for specialities such as *bacalao al pil pil* (salt cod with chilli) and *polbo á feira* (octopus with paprika) followed by *torta de Santiago* (almond cake).

Over a quarter of the pilgrims who register to walk the route start at the town of Sarria in Galicia, 116km (72 miles) east of the shrine, and complete the walk in five or six days. Accessible by railway and road, Sarria is a convenient choice. Five longer caminos – the Francés, Primitivo, Norte, Interior Vasco-Riojano and Lebaniego routes – are UNESCO World Heritage listed, as is Santiago de Compostela's elegant historic centre. The Camino del Norte passes through Bilbao and Santander, both of which have international ferry ports. The most popular long-distance route is the 780-km (485-mile) Camino Francés from Saint-Jean-Pied-de-Port in the foothills of the French Pyrenees; some pilgrims precede this with a visit to the Catholic shrine in Lourdes, also in southwest France. Whichever route you choose, you'll be rewarded with rolling green scenery, the company of like-minded souls and the satisfaction of accomplishment.

ALSO TRY

Kumano Kodo, Japan: This ancient network of paths connecting Shinto shrines on the mountainous Kii Peninsula is UNESCO World Heritage listed. Believers would undertake rigorous religious rites as they travelled.

Via Francigena, Canterbury to Rome: Christian pilgrims have followed this route since at least the 7th century. Accommodation along the way is patchy, so it's best to join an organized trip or plan to camp.

Abraham Path, Middle East: This community-based tourism initiative retraces the travels of Abraham, patriarch of several faiths, through Turkey, Syria, Jordan, Palestine and Israel. It currently stretches over 2,000km (1,240 miles).

Opposite: whichever route you choose through northern Spain, you'll have the chance to enjoy peaceful countryside, untroubled by traffic. Below: a shell in the pavement, such as this one in Burgos, indicates you're on the right track.

Gallivant through Bavaria on the King Ludwig Way

SOUTHERN GERMANY

Immerse yourself in stories of royal benevolence and intrigue on a beautiful hike in the northern foothills of the Alps

Do Wagnerian heroes and fairy-tale castles fire your imagination? What about remote hilltop forests, wreathed in legends of deities and dragons? If so, you have something in common with Ludwig II, one of Germany's most enigmatic and eccentric rulers, who was king of Bavaria from 1864 until his mysterious death, aged 40, in 1886. A 120-km (75-mile) hiking trail, the King Ludwig Way, explores the heart of his kingdom, leading from Starnberg, near Munich, to Füssen on the edge of the Ammergau Alps, close to the Austrian border.

Sometimes called the Fairy Tale King, Ludwig II surrounded himself with theatricality. He was a generous patron of the arts: among the host of creatives he funded was the composer Richard Wagner, who completed *Der Ring des Nibelungen* with his support. Behind the scenes, Ludwig lived like a fairy-tale character – sleeping by day, staying awake all night, taking moonlit walks and sleigh rides, and preferring friendly chats with ordinary locals to courtly discourse.

It was Ludwig's taste for flamboyant architecture that led to his most visible legacy. Digging into his personal fortune, he commissioned several fantastical palaces including Neuschwanstein Castle (*pictured, left*) in southwest Bavaria, a hilltop confection stuffed with murals depicting Wagnerian scenes. Topped with Romanesque turrets, it inspired Walt Disney's Sleeping Beauty Castle, replicated in theme parks around the world.

On the trail of a royal legend

You will start your walk at the end of King Ludwig's tragic tale. The route, waymarked with signs bearing a small blue 'K' wearing a crown, begins at the Votivkapelle beside Lake Starnberg, around 30km (18 miles) southwest of Munich. This neo-Romanesque chapel was built to mark the place where the monarch and his physician Dr Godden were found drowned in 1886. From here, you'll continue southwest, following farmland and woodland tracks through Bavaria's gently rolling, bucolic scenery. Along the way, there's the splendid baroque architecture of Andechs Monastery and Marienmuenster Church to admire on the banks of Lake Ammer, along with pretty villages of Alpine chalets and churches with onion-domed towers. The highest point on the walk is the 988-m (3,241-ft) summit of Hohenpeissenberg, possibly the finest viewpoint in Bavaria.

THE LOWDOWN

Best time of year: April to October. July and August is peak season, as is late September and early October, when Munich's Oktoberfest is in full swing.

Plan your trip: Allow at least eight days, including six days to complete the 120-km (75-mile) trail from Starnberg, near Munich, to Füssen, walking 14–28km (9–17 miles) per day. Organized tours, including road transfers and accommodation at country hotels, are available.

Getting there: Munich is around 1,100km (700 miles) by road and sea from London via Kent and Belgium. The train journey from London St Pancras via Paris or Brussels and Frankfurt takes around ten hours. If you opt for the sleeper train from Brussels — a civilized way to travel — it's around 19 hours. Ferries from Dover, Harwich, Hull or Newcastle to France, Belgium or the Netherlands link to Germany by road and rail.

The King Ludwig Way
rarely strays far from
a village *Gaststätte*
(or inn), so you'll have
plenty of opportunities
to sample Bavarian
specialities such as
Wurst (sausage),
*Knoedel (*dumplings),
Sauerkraut (pickled
cabbage) and, of
course, locally
brewed beer.

IN THE KNOW

As well as visiting the
Votivkapelle dedicated
to Ludwig II (*pictured
right*), don't hesitate
to push on the doors
of any other churches
and chapels you pass
on the King Ludwig
Way: inside, they're as
sumptuous as jewel
boxes, with elaborate
devotional paintings,
rococo gilding, marbled
pillars and twinkling
chandeliers.

The walk ends in Füssen (*pictured, right*), a violin-making town with royal connections dating back over a thousand years. Ludwig spent a large part of his childhood in this part of Bavaria: his father King Maximilian II's summer residence was Hohenschwangau Castle, a neo-Gothic palace southeast of town, overlooking lake Alpsee. It was no coincidence then, that Ludwig chose a hilltop overlooking the River Pöllat, just east of Hohenschwangau, for his masterwork Neuschwanstein Castle, inspired by medieval legends of dragon-slaying knights. Sadly, he didn't live to see it fully completed.

ALSO TRY

Ammergau Alps, Bavaria: Germans love hiking so much that if you'd like to extend your walk, you're spoilt for choice. From Füssen, there are several one-day treks in unspoilt Alpine terrain. None require specialist mountaineering skills or equipment.

Palatinate Wine Trail, southwest Germany: Leading from Bockenheim near Frankfurt to Schweigen-Rechtenbach, this is an easy 11-day walk through a sunny wine region.

Painters' Way, Elbe Sandstone Mountains: Close to Germany's Czech border, this route is famous for its magnificent rock formations.

Cruise along the River Danube

EASTERN AND CENTRAL EUROPE

Waltz up or down a legendary European waterway on a modern, fuel-efficient boat

The Danube is majestic divide, flowing between the Alps and the Carpathian Mountains to separate the Balkans, Italy, southern Austria and southern Germany from northern Europe. Truly international, it passes through ten countries between its source in the Black Forest and its delta in the Black Sea. Along the way it links a splendid chain of cities including Vienna, Bratislava, Budapest and Belgrade.

Visiting these cities by river rather than by road or train offers a unique perspective on some of Europe's most glittering destinations. Every time you wake to a new view, you can ponder how life must have been for the fishermen, ferrymen and bargemen who have plied their trade on the Danube over the centuries.

For responsible travellers, cruising demands an element of research, as some ships can have a negative environmental impact. To enjoy the Danube in 21st-century style, look for a cruise company with modern, eco-friendly vessels and policies. The best operators have ships equipped with solar panels and hybrid engines, which reduce carbon emissions, pollution, vibration and noise.

There are cultural considerations, too. Those who like to immerse themselves in local culture may miss the enjoyment of staying in distinctive city hotels and eating at neighbourhood bistros. It's also worth remembering that you may not see much of the countryside between stops, as your ship will often travel by night.

However, there are compensations. Travelling in what is arguably a floating eco-hotel lends a thread of continuity to your experience, and ship life is undoubtedly convenient, comfortable and convivial. For those who want to squeeze the maximum number of cities into a limited time period, it makes good sense.

THE LOWDOWN

Best time of year: April to November.

Plan your trip: River cruise itineraries and dates are fixed. The starting points on offer include Amsterdam, Nuremberg, Passau, Budapest and Bucharest. Allow eight to ten days afloat for a short Danube itinerary, or up to 23 days for a longer journey including the Rhine and other waterways.

Getting there: Transport options to Amsterdam from the UK include Eurostar trains from London St Pancras (3hr 50min) and DFDS ferries from Newcastle (15hr 45min to IJmuiden). For other destinations, take trains via London and Paris, Brussels or Amsterdam, or travel to Dover, Harwich, Hull or Newcastle for a crossing to France, Belgium or the Netherlands, then continue by road or train.

The partially restored ruin of Aggstein Castle, founded in the 12th century, towers above the Danube in Austria's Wachau Valley.

A dance through time

Danube cruises often focus on a stretch of the river that can be explored in eight days or so – Budapest (*pictured, right*) to Nuremberg or Passau, say. But it's also possible to travel all the way from Bucharest (near the delta) to southern Germany (near the source), and then continue cruising northwest to Amsterdam via the Main, Rhine and Waal. A journey such as this, taking over three weeks, allows you to savour Europe on a continental scale, relishing the changes in scenery and atmosphere as the miles roll by, and dosing up on history in the on-board library. A fine way to begin is to train-hop from the UK or western Europe to Romania on Europe's comfortable sleeper services. This is an unforgettable long-distance rail route that reveals the Danube and its cities from a different perspective. A journey combining the Brussels to Vienna Nightjet sleeper with the slightly longer leg from Vienna to Bucharest can take as little as 48 hours, including a day in Vienna.

During your time on land, you could take a waltzing masterclass in Vienna, visit an Art Nouveau spa in Budapest or visit one of Belgrade's many museums. While blockbuster stops such as these won't disappoint, passengers are often pleasantly surprised by the Danube's lesser-known landmarks. UNESCO World Heritage listed Regensburg in Bavaria, for example, is one of Europe's best-preserved medieval cities and breathtakingly picturesque, while the Golubac Fortress in Serbia (*pictured, below*) has spectacular river views from its newly restored towers.

ALSO TRY

River Douro, Portugal:
A voyage on an energy-efficient river cruiser, through the characterful city of Porto and one of Europe's oldest and greatest wine regions.

River Elbe, Germany and Czech Republic:
Discover castles, galleries and historic sites as you cruise from Berlin to Prague.

Stroll through Slow Food country in Piedmont

NORTHWEST ITALY

THE LOWDOWN

Best time of year:
May to December.
Alba's Fiera del Tartufo
(white truffle fair, with
stalls and tastings) is
held every weekend
from October to
early December.

Plan your trip: Guided
trips are typically seven
nights long and include
accommodation and
restaurant bookings.
Bra and Alba both make
good bases. Linked by
direct trains, they're
16km (10 miles) apart,
southeast of Turin.

Getting there: Bra is
around 1,300km
(800 miles) by road
and sea from London
via Kent and France.
Rail routes from the
UK typically begin
with a leg to London
St Pancras. From
here, trains to Bra
via Paris Nord, Paris
Lyon and Turin take
around 10hr 50min.
Alternatively, travel to
Dover, Harwich, Hull or
Newcastle for a crossing
to France, Belgium or
the Netherlands, then
continue by road
or train.

Work up a healthy appetite on leisurely walks in the hills around Alba and Bra

Slow food: there was a time when those words meant, at best, a lazy, late lunch. All credit to the Italians for recognizing that our cooking and eating habits can be a force for good. In 1989, the Piedmontese gourmet Carlo Petrini and a group of fellow gastro-activists founded the Slow Food movement, which has since gone from strength to strength. Their chief aim is to defend and celebrate regional culinary traditions and the pleasure of living life at a gentle pace.

Visitors to Piedmont's hillside vineyards and hazelnut groves can immerse themselves in Slow Food culture on a guided or free-ranging walking tour, meandering along riverside paths and farmers' trails to hilltop villages, where sumptuous evening meals with quaffable local wine await.

Indulgent as this may sound, there's far more to Slow Food than harmless *ghittoneria*, or gluttony. Over the years, the movement has gone global, evolving to recognize the strong connections between plate, planet, people, politics and culture. It campaigns for agricultural biodiversity, supporting small-scale, sustainable food producers, and promotes culinary biodiversity through a project called The Ark of Taste, which catalogues endangered traditional foods. In doing so, it throws a spotlight on the knowledge and talents of indigenous people, highlights the relationship between food and climate change, and launches a many-pronged attack on hunger and malnutrition throughout the world.

Carlo Petrini's hometown of Bra, the proud birthplace of Slow Food, remains the movement's hub. Alba, capital of the Langhe, Piedmont's celebrated UNESCO Human Heritage winemaking region, is equally committed to the cause. You'll discover these places, and more, as you eat your way around Piedmont, a region famous for red wine, white truffles, cheese and beef.

Surrounded by vines, Castiglione Falletto is one of a chain of hilltop villages within Piedmont's UNESCO World Heritage listed Vineyard Landscapes.

Food for the soul

To keep things simple, organized Slow Food tours use a comfortable hotel as a base, transporting groups to the start of each walk by minibus. On some days, you may be offered a choice of routes, one easier and the other more strenuous, and visits to wineries, cheesemakers or *grissini* bakers could be built into the programme. Lunch might be a picnic of bread, cheese, ham, pastries and fruit, purchased at village shops during your walk; and come evening, there will be a suitably celebratory dinner, featuring wine from one of the vineyards you ambled through during the day.

This is a region of life-affirming views, with church towers soaring above terracotta roofs and neatly planted fields spread like pleated skirts on the slopes below. As appealing to photographers as it is to foodies, it's particularly beautiful in the golden light that drenches the hilltops towards the close of day.

The 14th-century castle that presides over Grinzane Cavour is home to a Michelin-starred restaurant, serving beautifully presented regional specialities.

ALSO TRY

Tuscany: Join a gastronomic tour, meeting artisan producers of balsamic vinegar, hand-rolled *pici* pasta and *cavolo nero* (Tuscan kale), an essential ingredient in minestrone.

Chianti: Set off on a wine-focused walking trip, following woodland tracks from one charming village to the next, tasting the best of the local vintages.

Puglia: Enjoy a farmhouse cooking holiday, learning how to rustle up seafood risotto and *orecchiette con cima di rapa* (pasta with turnip-top greens) with all the flair of a local.

Ride an electric bike through the Tuscan hills

CENTRAL ITALY

Dip into la dolce vita on an easy-going pedal through quintessentially Italian landscapes

Think of Tuscany, and it's likely to conjure up an image of a vine-clad hill, topped by a honey-stoned *castello* and a grove of cypress trees. Wedged between the rugged Apennine Mountains and the Tyrrhenian Sea, Tuscany's elegant contours – and rural traditions – are integral to its charms. Its thriving *agriturismo* network of working farms with comfortable beds, hearty home cooking and stunning views, makes this an appealing region to explore.

Tuscany is tremendously popular with holidaymakers. But if you're fit and keen, a walking or cycling trip along its winding lanes offers that little bit extra – the chance to venture into areas where day-trippers, nipping about in their rented Fiat 500s, rarely reach.

If you're worried about those contours, bringing or renting an electric bike (or, for a guided experience, joining an e-biking tour) is the way to go. You don't necessarily have to ride the whole way: it's possible to carry e-bikes on some European ferries, buses and trains (the rules vary, however, so check in advance). E-bikes are not just for softies – athletic types can dial up the resistance to give their legs a proper workout – but for those who'd like a little boost on some of the climbs, they're ideal. Pedalling uphill on electric power, it feels as if an invisible giant's hand is pushing you gently from behind.

E-biking may be a piece of cake, but it's still good exercise: so when lunchtime rolls around and the table is loaded with *panzanella* salad, spinach-and-ricotta ravioli and juicy wood-fired pizza, you can relax, and dig in with glee.

Best time of year: May to June or September. Avoid July to August, when the weather can be uncomfortably hot.

Plan your trip: Allow five to ten days for a beginners' trip, cycling for up to three hours each day. Options include guided or self-guided tours using luggage transfer services, starting from Florence, Pisa or Cortona. Alternatively, travel independently.

Getting there: Tuscany is around 1,600km (1,000 miles) by road and sea from London via France and northern Italy. Eurostar trains from London St Pancras connect with trains from Paris to Florence and Pisa via Turin or Milan, taking around 13 hours (or 24 hours if you opt for an overnight stop). Ferries from Dover, Harwich, Hull and Newcastle connect by road and rail to Italy.

CASTAGNACCIO ALLA TOSCANO

If you're in Tuscany in autumn, be sure to try a slice of *castagnaccio*, a simple, seasonal chestnut dessert made with chestnut flour, sultanas, pine nuts, olive oil and rosemary.

ALSO TRY

Southern Italy: E-bike across the 'ankle' of Italy from the Adriatic coast to Sorrento, via the strikingly picturesque Lucan Dolomites and Padula's palatial monastery.

Costa Brava: Join an e-biking tour of the rugged Catalonian landscapes that inspired Salvador Dalí, Joan Miró and Pablo Picasso.

Explore the Skåneleden on foot

SOUTHERN SWEDEN

Skåne's well-signposted hiking trails make short walks and longer trips a breeze: just grab your backpack and go

The beautiful countryside of Skåne, the southernmost region in Sweden, is a walker's dream. The Skåneleden, a collection of waymarked long-distance walking routes, radiates into all four corners of the region from the friendly, rejuvenated city of Malmö. Winding along quiet lakeshores, through woodlands and past impressive ravines, these paths are among Europe's best.

The trail divides neatly into six regional sub-trails, the newest of which, SL6 (Vattenriket, or Kingdom of Water), will be complete in 2023. You could choose one or more sub-trails, or opt for one of the 105 day-walks they contain. If you were to walk the lot, you'd clock up around 1,250km (777 miles) – roughly the distance from Malmö to Stockholm and back.

A section of sub-trail SL 5 (Öresund) called the Kullaleden, a 70-km (43-mile) coastal loop around the Kullen peninsula from Utvälinge to the medieval city of Helsingborg, is particularly famous. Classed a 'Leading Quality Trail' by the European Ramblers Association for its beautiful setting and user-friendly infrastructure, it's an invigorating three-to-five-day walk, following the rocky shoreline of the Kattegat sea, through traditional villages with tidy little harbours and beaches backed by farmland. Each walk's relative difficulty is graded: while some have long, steep ascents, others are level, smooth and short enough for families with young kids to tackle with ease. One 30-km (19-mile) chunk of the Kullaleden, the Kulla-Rulla, has been adapted for wheelchairs and pushchairs, and it's hoped that additional stretches of the Skåneleden will be fully accessible in the future.

To help you navigate, download the Skaneleden smartphone app, Vandra i Skane (*hiking in Skane*), which has route maps for selected trails overlaid on satellite imagery. It also includes notes on 200 points of interest, over 600 minutes of stories recorded in Swedish and English for the vision-impaired, and sign-language videos for the hearing impaired.

Perfect for paddling and swimming, Helsingborg's Fria Bad is a beach to make you feel like a kid again.

SKÅNELEDEN
SUB-TRAILS:

SL1 (Coast to Coast),
370km (230 miles),
26 days

SL2 (North to South),
325km (202 miles),
28 days

SL3 (Ridge to Ridge),
162km (100 miles),
14 days

SL4 (Österlen),
188km (117 miles),
14 days

SL5 (Öresund),
214km (133 miles),
18 days

SL6 (Kingdom of Water),
50km (31 miles),
5 days

Ancient forests and hidden villages

The trail offers a satisfying choice of landscapes. To enjoy the leafy wonderland that is Söderåsen National Park (*pictured, below*) – the largest unbroken area of protected woodland in northern Europe – take the train to Åstorp and head southeast along sub-trail SL3 (Ridge to Ridge).

Alternatively, to bury yourself in rural Sweden, head for Osby and pick up sub-trail SL1 (Coast to Coast) as it winds east to the village of Verum on the meandering Vieån River. On the way there, you'll stroll past birch trees and meadows of hare's-tail cottongrass and cloudberries, then deep-dive into spruce forests. Abba's Agnetha and Björn were married in Verum's medieval church in 1971.

Malmö, a pioneer in sustainable urban regeneration, makes a worthwhile stopover before or after your walk. Its strikingly renovated Western Harbour, home to Santiago Calatrava's landmark Turning Torso building (*pictured, right*) and flats powered by renewable energy, has fine views of the Öresund Bridge, gateway to Copenhagen. Multicultural and cycle-friendly, the city has much to offer, including open-air swimming at the Kallis seawater pools and eclectic shopping at vintage markets.

PACK A TENT OR TRAVEL LIGHT

Camping is the best way to immerse yourself in nature on the Skåneleden. To lighten your load on the trail, you could book a baggage transport service. Alternatively, pick a route with bothies and guesthouses nearby, such as SL4 (Österlen). Or choose day hikes: some are less than 6km (4 miles), so all you'll need is a water bottle (and perhaps your swimming things, ready for a dip in a lake).

ALSO TRY

Karhunkierros Trail, Finland: This 82-km (51-mile) route explores a region of silent pine forests and rushing waterfalls, rich in unique flora and fauna.

Laugavegur Trail, Iceland: World famous for its dramatic mountain landscapes under big skies, this 55-km (34-mile) trail takes three to five days.

La Gomera, Canary Islands: Follow rugged paths through primeval laurel forests, draped with moss and ferns.

Wend your way through southern Transylvania by bike

CENTRAL ROMANIA

Let time slow down as you pedal past wildflower meadows and medieval fortified churches in unspoilt Saxon villages

It's easy to blame Bram Stoker for Transylvania's sinister reputation. Inspired by tales of Romania's Vlad the Impaler, he decided a Gothic castle in the Carpathian Mountains would make a perfect palace for his fearsome fictional creation, Count Dracula. The region has been synonymous with vampires and horror stories ever since. Why would you choose to go cycling here? And if you did, should you pack your panniers with garlic?

Hold the garlic and pack your camera instead. In reality, the hills and valleys of south Transylvania are blissfully bucolic, with tranquil pastures, wildflower meadows and beech forests. That's not to say they're hazard-free – villagers occasionally see bears cross their backyards, and hear wolves howling at night – but the chances of a perilous encounter are reassuringly slim.

The counties of Mureș, Sibiu and Brașov are dotted with UNESCO World Heritage listed villages founded by Transylvanian Saxons: wealthy and influential German farmers, artisans and merchants who migrated southeast to settle here between the 12th and 16th centuries. Threatened by the Ottomans and Tatars, they built their villages around Gothic and Romanesque churches, fortified with chunky curtain walls and towers. Over 160 of these fortress-like churches still dominate Transylvania's landscapes today.

THE LOWDOWN

Best time of year: April to June and September to October. In July and August, temperatures can top 30°C (86°F).

Plan your trip: Organized tours are available, cycling 18–30km (11–19 miles) a day for five days with a guide and support team to transport your luggage. It's also possible to travel independently. Brașov, the regional hub, has good train and bus connections within Romania and to Austria and Hungary.

Getting there: Brașov is around 2,420km (1,500 miles) by road and sea from London via Belgium, Germany, Austria and Hungary. Trains from London St Pancras via Brussels take around 33 hours, including a night in Frankfurt and a sleeper from Budapest to Brașov. Romanian bus companies run direct from Birmingham, Northampton and other UK cities (36hr). Ferries from Dover, Harwich, Hull and Newcastle to link to routes to Romania by road or train.

Backed by the Făgăraș Mountains, the village of Hosman is dominated by a fortified Romanesque church where a music festival, Holzstock, takes place each August.

VLAD THE IMPALER

Bram Stoker may have invented Dracula, but he didn't create the name: it belonged to Romanian national hero Prince Vlad III (aka Dracula), the violent son of Vlad II Dracul of Wallachia. He was born in Sighișoara, which is now a UNESCO World Heritage Site.

IN THE KNOW

To dig deeper into Saxon culture, spend a few days in one of the villages, perhaps going on horse and cart excursions to visit local craftspeople, or heading into the meadows and forests to learn about wildflowers and mushrooms.

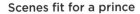

Scenes fit for a prince

With little traffic and relatively few visitors, the Carpathian foothills are a great place to ride. Pedal along cart tracks and country lanes on a bike (or e-bike) with a guide, and you'll have opportunities to chat with locals, many of whom still speak a distinctive German dialect.

Zigzagging from village to village via fields, oak woods and orchid meadows, you'll make regular stops for regional cuisine such as roasted meat with organic pickles and *székely köménynes* (caraway brandy).

On an easy-going itinerary, there'll be time to visit the pretty town of Sighișoara and fortified churches such as Biserica din Mălâncrav, with its medieval frescoes depicting biblical scenes. Lutherans whitewashed these superb paintings into obscurity during the Protestant Reformation, but they were rediscovered during restoration works in the 20th century.

You'll stay in village guesthouses, which might be anything from a 15th century mansion or Gothic townhouse to a charming Saxon cottage with a terracotta-tiled roof and flower-filled orchard, or even a homestead owned by a member of the British royal family.

Prince Charles believes passionately that Transylvania is one of the last corners of Europe where people live sustainably, in harmony with nature. To help protect local lifestyles and landscapes, he bought an 18th-century homestead in the Zalán Valley and commissioned its restoration and conversion into guest accommodation, creating much-needed part-time jobs for local farmers. Full of antique tiles and furniture, it provides modern comforts in a genuinely historic setting.

ALSO TRY

Țarcu Mountains: Track wild bison, Europe's largest land mammal, in the Southern Carpathians, southwest Romania. Herds were reintroduced in 2014 by Rewilding Europe and WWF Romania.

Piatra Craiului National Park: Go bear-watching. A 5,000-strong population of European brown bears can be found among the spruce trees in this park near Brașov, central Romania, along with eagle owls, wolves, deer and bats.

Paragliding over the Carpathians: Brașov is surrounded by great launch points. Take a tandem flight with an instructor, for unforgettable views of the forested slopes.

To dial the excitement up a few notches, cyclists can tackle Transylvania's famous Transfăgărășan, a mountain road that hairpins through the Southern Carpathians.

Paddle the River Soča

THE LOWDOWN

Best time of year: April to August. The kayaking and rafting season runs from 15 March to 31 October. Conditions vary with the volume of melting snow and rain.

Plan your trip: Allow at least three days in the Soča (pronounced *sotcha*) Valley, perhaps as part of a road trip around Slovenia.

Getting there: Bovec is around 1,510km (940 miles) by road and sea from London via France or Belgium and central Europe. Driving your own kayaking gear from the Dover ferry or Folkestone Eurotunnel to Bovec takes around 16 hours, plus stopovers. The nearest convenient station is Arnoldstein in Austria. Trains from London St Pancras via Brussels, Cologne and Villach, followed by a 50-minute cab ride to Bovec, take around 18 hours.

With beginner-friendly shallows and gnarly rapids, the beautiful Soča Valley ranks among Europe's finest river-kayaking spots

Few European rivers are as entrancing a colour as the Soča. On sunny days, it looks turquoise or even emerald green, thanks to its crystal-clear water and bone-white riverbed. Restless rapids marble the surface with foamy white. Backed by the crinkle-capped Julian Alps in the Triglav National Park, it's a sight to move romantic souls to poetry, local greats Simon Gregorčič and Giuseppe Ungaretti among them.

Gushing steeply downhill from Trenta to the Italian Adriatic, the Soča is lively enough to impress adventure-seekers, too. Compared to other Alpine whitewater kayaking and rafting destinations, it's still relatively unknown, and all the more exciting for that.

Conditions can be challenging, so it's best to have a grasp of the basics from the outset. Book an intermediate kayaking course at one of the activity centres in Bovec, hub of the Soča Valley, and you'll learn to optimize your strokes among the Soča's rocks and eddies, focusing on hydrodynamics, whitewater reading skills, advanced steering techniques and edging (quickly tilting your boat, in order to turn rapidly and under control).

The water between Čezsoča and Boka near Bovec is graded I–II: wide, smooth and gentle, it's perfect for paddling practice. Downstream at Most na Soči, the Soča widens even further into a turquoise reservoir, with terracotta-roofed houses perched on the bank.

If you find yourself longing to crank up the pace from slow to supercharged, simply head for the stretches where the water squeezes through narrow ravines and bubbles over submerged rocks. The tricky Koritnica, a tributary of the Soča east of Bovec, and the grade IV–VI rapids between Trnovo and Otona will really put you through your paces.

SAFETY ON THE SOČA

To kayak the River Soča, you need a Soča Valley pass, available by the day, week or season at local activity centres; cold-water gear, a helmet, rescue rope and a firstaid kit are recommended. You must only enter or exit the river at official access points, and all but the most experienced are advised to be accompanied by a guide.

ALSO TRY

Gorges de l'Ardèche, France: Enjoy a gentle paddle along a 30-km (19-mile) route, passing under its famous natural stone bridge, the Pont d'Arc.

Tara Canyon, Bosnia: This UNESCO World Heritage listed canyon is Europe's largest, with grade III–V rapids. Head here for high-energy kayaking.

Feel on top of the world in the Atlas Mountains

MOROCCO

Hike across the roof of North Africa with Morocco's highland-dwelling Amazigh people

South of the red-gold city of Marrakech, the burnt umber peaks of the Atlas Mountains mark the transition between Mediterranean North Africa and the seven-million-year-old Sahara. This is prime hiking territory, where you'll find hillsides as empty as Marrakech is frenetic.

With routes ranging from multi-day hikes to short mountain walks that you could complete on a day-trip from town, there's something for everyone in the Atlas. To make the most of these inspiring surroundings, all you need are a decent level of fitness and a willingness to rough it. Most tracks are rough and rocky, reaching altitudes exceeding 1,800m (6,000ft) – considerably higher than the UK's tallest mountains. On a trek lasting several days, you'll be staying in basic village accommodation, so to smooth your path, it's best to hire a team to guide, cook and carry, with mules to assist.

The local Berber people, or Amazigh (pronounced *ama-zeer*) as they prefer to be known, are famously hospitable, with a talent for poetry, chanting and mesmerizingly rhythmic music played on the *gimbri* (a three-stringed lute) and *qraqrabs* (metal castanets). Throughout North Africa, they struggle for official recognition of their linguistic and cultural traditions. It's a privilege to spend time with them, and perhaps learn a little about life in this challenging terrain.

Exploring the rugged uplands on foot

Toubkal National Park is the most popular destination, with good reason. Some 70km (43 miles) south of Marrakech, its clear mountain air comes as a tonic after the heat and noise of the city. Set in the High Atlas – the range's lofty core – the crinkled terrain of Toubkal Massif fans out around the snow-capped peak, Mount Toubkal. At 4,167m (13,671ft) it is the highest mountain in North Africa. Daunting as that may sound, it's a non-technical climb that's perfectly possible for hikers with good altitude tolerance. Starting from the village of Imlil, you could make it to the top in a day, but it's wiser to spend a night at a Toubkal refuge, acclimatizing, before pushing on to the peak in the clear light of morning.

THE LOWDOWN

Best time of year: March to October. In winter, nights can be cold. There's usually snow on the mountains from mid-January to mid-February.

Plan your trip: Allow at least ten days for a return trip. Organized guided walking tours are available. It's also possible to travel independently. If you're climbing Mount Toubkal, you're required to hire a local guide. On multi-day hikes, porters are a good investment.

Getting there: Imlil, gateway to Toubkal National Park, is around 3,100km (1,930 miles) by road and sea from London via France and Spain. It's possible to travel from London by train to Algeciras, bus to Tarifa, ferry to Tangier and train to Marrakech (see page 66), then two-hour bus or taxi ride south to Imlil. This takes around 31 hours, plus stopovers. Ferries from Portsmouth and Plymouth to Spain connect with trains to Algeciras.

Imlil's celebrated Kasbah du Toubkal, a mountain guesthouse with glorious views, is a great starting point for hikes, even if you're staying elsewhere. Run in partnership with the local community, it raises funds for local development projects. Its satellite guesthouse, Azzaden Trekking Lodge, is several steps more comfortable than the rustic hostels that are the norm in these parts.

As you hike across the Azzaden Valley, you could visit villages such as Tizi iZggar, drinking mint tea with local families or visiting rural education centres. The steep, open landscapes enhance a sense of immersion in Amazigh life, with houses seemingly piled one on top of another, immaculately dressed women tending their crops and mules trotting along the dusty tracks.

IN THE KNOW

The muleteer carrying your luggage will often follow a different route to you, so you won't have access to your stuff during the day. Bring mini bottles and packages of sunblock, lip salve, sanitizer and tissues to keep your daypack light, and don't forget your refillable water bottle.

ALSO TRY

Jordan Trail:
Spectacularly ambitious, this 650-km (400-mile) route opens the entire country of Jordan up to hikers. Highlights include Petra, Wadi Rum and the Dead Sea canyons.

Balcony Trail, Oman:
Follow a waymarked route along the ridges of the Hajar Mountains, through arid highlands that are as dramatic as any in the Middle East.

EPIC VOYAGES

It's green to be a minimalist. But when it's time to plan your next travel adventure, why not think big? Crossing continents by train, bus or car is an exhilarating way to see the world, absorbing all the fascinating little details that you'd miss if you flew. Eventually, you'll reach the ocean, but that's no reason to stop: unforgettable journeys by yacht, ferry and ship await.

Cross Europe by bus, from Paris to Kiev

FRANCE AND UKRAINE

Travel to one of Eastern Europe's oldest cities, for sumptuous churches, spacious squares and cool nightlife

Long-distance bus versus train: which wins? In some parts of the world, there's no contest. In South America, for example, the rail network is fragmented and intercity buses rule. But in Europe, where both trains and buses are widely available, the choice isn't as clear-cut. On the journey from France to Ukraine, it could go either way, but for convenience, the bus wins out.

Comfort-wise, modern European long-distance buses and trains are much of a muchness, with quality seats the norm, and WiFi and power sockets increasingly common. A key distinction comes when considering overnight services: sleeper trains usually offer a better night's rest than buses, even though some buses are adapted for long journeys, with reclining seats and extra leg room. Trains also tend to be much faster and, scenery-wise, they have the edge, offering views of remote countryside unspoilt by tarmac, cars and lorries. On popular routes, booked well in advance, bus and train fares are similar: price comparison websites can show you this at a glance. However, on less commonplace routes, or during peak times, or at short notice, buses can be considerably cheaper.

If all other things are equal, network connections may clinch your choice. Europe's low-cost and independent international bus network is extensive, covering well over 2,500 destinations, with more direct services than the railways. Long-distance routes include regular pauses and brief stops at the end of each driver's shift, but you won't necessarily have to change buses.

The 2,400-km (1,491-mile) journey from Paris to Kiev takes a similar length of time by train or bus – between 40 and 48 hours. But whereas the train journey has confusing permutations requiring at least two changes and three tickets, booked separately, the bus journey is simple, with direct services often available. Typically, bus routes proceed due east via Reims and Metz in France, Frankfurt and Dresden in Germany, Krakow in Poland and Lviv in Ukraine, with a brief stop in southern Poland. Buses that depart in the morning offer fleeting views of swathes of northern France and Germany before night falls. On your second day, you'll be speeding through Poland and west Ukraine, where Lviv's beautiful, medieval centre is worth a visit in its own right.

Kiev's flamboyant 18th century Church of St Andrew was the work of famous Italian architect Bartolomeo Rastrelli, a master of the baroque.

THE LOWDOWN

Best time of year: April to June and September to October. Summer and winter can bring extreme temperatures.

Plan your trip: Allow at least ten days for a return trip with a few days in Ukraine; longer, for more time exploring Eastern Europe.

Getting there: Paris and Kiev are around 460km (285 miles) and 2,500km (1,500 miles) respectively by road and sea from London. Eurostar trains from London to Paris take 2hr 15min. Bus routes from the UK to Paris typically begin with a leg to London Victoria; from here, direct services take 8hr 45min. Alternatively, travel to Portsmouth, Newhaven, Dover, Harwich, Hull or Newcastle for a ferry to France, Belgium or the Netherlands with links to Paris. From Kiev to the UK, options include buses and trains via Poland, Germany and Belgium.

ANNE DE KIEV

The cultural connections between France and Ukraine date back to medieval times. In 1051, King Henry I of France married a Ukrainian, Anne de Kiev, in Reims Cathedral. Anne was well educated and multi-lingual at a time when many people, including her royal husband, could not even sign their own name. She is credited with bringing many innovations to French society, from bath-houses to cutlery. Her Cyrillic bible, a gift from her father, came to be known as the Reims Gospel, and was the book on which subsequent French kings took their coronation oath.

Haze blurs the skyline beyond Kiev Pechersk Lavra (the Kiev Monastery of the Caves).

Kiev and beyond

Elegant and cultured, but far less visited than Paris, the Ukrainian capital is a city of gilded cathedrals and temples to culture, such as the National Opera and the open-air National Museum of Folk Architecture. It also has a buzzing club scene – Kiev has been called Berlin's cheaper but cooler cousin.

Out of town, the curious could visit Chernobyl, to learn about its 1986 nuclear disaster, or could arrange a homestay on a farm, perhaps in the Carpathian Mountains. Ukraine's well-established Association for the Promotion of Rural Green Tourism connects homestay hosts with tourists who'd like to immerse themselves in pastoral life. Communicating in the universal language of smiles and gestures, guests can expect a clean bed, generous quantities of homecooked organic *borscht* (beetroot soup), *pampushky* (garlic doughnuts) and *salo* (cured pork fat), and access to beautiful hiking trails.

Ferris wheel at the abandoned Pripyat amusement park near Chernobyl. Five days before the park's grand opening was due, the world's worst nuclear disaster occurred.

IN THE KNOW

Compared to western European cities, Kiev can feel polluted. If you're in need of a breather, head south to Holosiivskyi National Nature Park, a swathe of forests, lakes and footpaths.

ALSO TRY

A hop-on-hop-off bus trip: Busabout's backpacker-friendly European routes include Paris to Switzerland, Munich, Austria, Poland, Hungary and the Balkans.

Europe to Morocco by bus: This is the furthest south you can reach on a scheduled European bus service. The journey from London to Marrakech, for example, takes around two and a half days via Spain and the Strait of Gibraltar.

Eight time zones without flying: London to Singapore

EUROPE AND ASIA

Whether you ride it, race it or take it at a leisurely saunter, this is one of the world's most illuminating long-distance routes

Ultra-modern Singapore is so far from London that it's barely possible to see both destinations on a globe without spinning it. Ever since the 1930s, it's been possible to fly between them; long-range aircraft don't even need to touch down en route. At 10,849km (6,741 miles) and just over 13 hours, this is the longest nonstop flight between European and Asian capitals that commercial airlines offer.

What if you were to travel overland instead? Singapore is the most distant city you can reach from western Europe without leaving dry land, and to get there, you must cross a couple of dozen borders. The prospect of doing so without flying has caught travellers' imaginations for generations: taking you through parts of central Asia that few tourists ever see, it's one of the world's classic responsible tourism adventures. It's in this spirit that the UK's BBC chose London to Singapore for its first *Race Across The World* challenge, creating a hit 2019 television show that opened many British viewers' eyes to the possibilities presented by fully immersive, flight-free travel.

By road, the most direct route from London to Singapore is a mighty 14,670km (9,115 miles), requiring around 175 hours behind the wheel. In practice, danger spots and closed borders force overlanders to take a more circuitous route, typically looping through China to avoid Afghanistan and Pakistan. As an alternative to driving (or having people drive you), you could use ferries, trains, sleeper buses, cars and other vehicles. Some combinations are inevitably greener, faster or cheaper than others, and the scenery and stops vary a great deal, too.

For the ultimate low-impact adventure, you could cycle, if you have a year or so to spare. Many people have completed this impressive feat, either for charity or just for the hell of it, braving heatwaves, hailstorms and headwinds, but enjoying superb scenery, heart-warming encounters and the intoxicating freedom of the open road. For some, it's been a life-defining experience.

THE LOWDOWN

Best time of year: Any time, however winter (November to March) brings bitter conditions to central Asia, typhoons can hit Southeast Asia between August and October and Singapore is rainiest from November to January.

Plan your trip: For a one-way journey by public transport followed by at least a few days in Singapore, allow a minimum of three weeks. If driving, allow at least a month. Cyclists typically take around a year to complete the route.

Getting there: London, the UK's main transport hub, has direct rail and road connections from most parts of the country. Singapore has one of the world's busiest container ports. For a flight-free return journey by a different route, it's sometimes possible to book passenger berths on cargo ships sailing to Europe via the Indian Ocean and Suez Canal.

VISIT AN ELEPHANT REFUGE

There's been a great deal of controversy about elephant exploitation in Southeast Asia in recent years. Refuges and sanctuaries such as MandaLao in Laos, the Elephant Valley Project in Cambodia and Mahouts Elephant Foundation in Thailand are trying to put things right by rehabilitating elephants that have suffered abuse. They invite visitors to meet and learn about elephants, and have a 'no rides, no tricks, no chains' policy.

The Mediterranean route

The TV-show race contenders were given fifty days and a cash budget equivalent to a full-price airfare to cover over 19,000km (11,806 miles) overland via fixed points in Greece, Azerbaijan, Uzbekistan, China and Cambodia. They made occasional stopovers to go sightseeing, hang out with locals or take on odd jobs to boost their funds, which had to cover food and accommodation as well as transport.

To reach Greece, some travelled overland via the Balkans; others made for Italy, then took a ferry from Venice, Bari or Brindisi. The speediest and smoothest route to Greece, taking little over 48 hours, is via train to Bari (Eurostar to Paris, TGV to Milan, Frecciabianca along the Adriatic coast) and overnight ferry to Patras. Continuing east, the contenders travelled overland to Istanbul, or took a ferry to southwest Turkey. For the massive journey from Turkey to Vietnam, they took buses to Baku in Azerbaijan, a ferry across the Caspian Sea and taxis, trains and buses across Kazakhstan, Uzbekistan and China, taking time to appreciate the region's striking desert landscapes, legendary, jewel-box cities such as Samarkand, and to visit the Great Wall of China.

Travelling overland through Vietnam immerses you in the details of daily life, from catching local transport to buying from street sellers.

IN THE KNOW

One of the most scenic and romantic journeys you can make in Southeast Asia is by boat along the Mekong River through Laos, Thailand, Cambodia and Vietnam, passing floating villages and markets, rice paddies, pagodas and temples.

GARDENS BY THE BAY

Singapore's top attraction spotlights the city-state's green principles. It's a high-tech botanical garden, with open parkland and energy-efficient, water-wise conservatories where you can stroll through Mediterranean meadows and a tropical cloud forest. Make sure you stay until nightfall, when solar-powered lights bathe the iconic Supertree Grove in glorious colours.

Twice a day, a passenger train rattles along a narrow lane in Hoàn Kiếm, Hanoi's hectic Old Quarter, hooting as it goes.

The fast-track route across Russia and China

To maximize your time in Asia, you could start by zooming from London to Vietnam by train. Travelling without stopovers, you could make it to Hanoi or Ho Chi Minh City in ten or eleven days. The fastest route is via Brussels, Moscow and the Trans-Siberian and Trans-Mongolian or Trans-Manchurian railways to Beijing (see page 204), followed by the 37-hour Beijing-to-Hanoi sleeper train. As an alternative, there's a geographically more direct railroute from Moscow to Beijing via Kazakhstan; however, this takes longer, and there's more red tape involved.

Travelling flat out by bus and train from Hanoi, you could reach Singapore within three days. But there are, of course, dozens of good reasons to move through Southeast Asia at a leisurely pace, exploring all the sights, sounds and flavours that Vietnam, Laos, Cambodia, Thailand and Malaysia have to offer.

ALSO TRY

Cape Nordkinn to Punta de Tarifa: If you like the sound of a trip between mainland Europe's northernmost and southernmost points, the drive from Mehamn in Norway to Isla de Tarifa in Spain is for you. It's a fairly direct 5,600km (3,480 miles): jump in an electric car, and you could cover it in around sixty hours, plus stops.

Historic Route 66, USA: The classic American road trip, this 3,940-km (2,448-mile) journey starts in Chicago and sweeps you west through St Louis, Oklahoma and Albuquerque, all the way to Santa Monica, California.

PERM, CITY OF CULTURE

Deep in the Urals, before
European Russia gives
way to Siberia, the
Trans-Siberian Railway
stops at Perm, one of
Russia's great cultural
centres outside Moscow
and St Petersburg.
Alight here for opera,
ballet and PERMM,
the city's museum of
contemporary art.

Take the Trans-Siberian Railway to Tokyo

RUSSIA AND JAPAN

Top off the world's longest train journey in style with a trip to Japan's green-thinking capital

THE LOWDOWN

Best time of year: Any time. The Trans-Siberian Railway is at a similar latitude to Scotland, with long daylight hours from May to August, but the climate is continental, with snowy winters. Japan's rainy season (June and July) is hot and humid.

Plan your trip: Train journeys from Moscow to Vladivostok include up to eight nights on board. Allow at least three days to continue to Tokyo by sea and rail or bus, including two to three nights on the weekly ferry from Vladivostok to Sakaiminato.

Getting there: Moscow and Tokyo are around 2,900km (1,800 miles) and 13,000km (8,000 miles) respectively from London by road and sea. Trains from London to Moscow via Belgium and Germany take around 48 hours, including a night in Brussels and a sleeper from Berlin. Moscow's international connections allow many other options. To return from Tokyo to London by sea, take cargo ships via China or Singapore.

If you enjoy ticking off record-breaking journeys and destinations – the highest this, the deepest that – a ride on the legendary Trans-Siberian Railway from Moscow to Vladivostok is a must. With a track measuring 9,259km (5,753 miles), it's the longest continuous passenger railway in the world.

Excitingly, it's also a corridor between cultures and continents. Hop off at Vladivostok, jump on a ferry across the Sea of Japan, and you can be in Japan within 44 hours. From your arrival point, the port of Sakaiminato in southern Honshu, the train to Tokyo takes just over seven hours.

Safe, comfortable and affordable, the Trans-Siberian Railway is a lifeline for a varied cross-section of travellers, from soldiers on leave, to students travelling between home and university. Several passenger trains, offering differing levels of service, use this route across the Urals and Siberia. The friendly and well-kept Rossiya trains – proudly decked out in the colours of the Russian flag – are considered the best option.

In summer, the most popular season for tourists to make the journey, the days are long and light; in winter, the snow-swept landscapes can be austerely beautiful. But you don't take this trip for the views, which are mostly of endless birch forests. Nor is it a luxury experience. If you don't mind sharing, there's no real value in splashing out on first-class tickets (second-class compartments sleep four rather than two). If you're sociably inclined, you may even prefer the dormitory-like *platzkart* third-class carriages.

Slow travel, Russian style

With an average speed of 90km/h (56mph), this is no whistlestop journey. There's a sense of togetherness among the passengers, many of whom bring stashes of black bread, cured meat, pickles, teabags and vodka, ready to travel long distances at a stretch. Others buy smoked fish and blueberry waffles from vendors on the platforms, or tuck into *pelmeni* (dumplings) and herring in the dining car.

IRKUTSK, PARIS OF SIBERIA

It was Chekhov who compared Irkutsk to Paris, and the nickname stuck. With historic timber houses and spacious neoclassical squares there's much to admire here, and Lake Baikal's waterside hiking trails are within easy reach.

PLASTIC-FREE TOKYO

Tokyo takes sustainability seriously. There are few rubbish bins in the streets, since most Tokyoites take everything home to sort and recycle, and food outlets use witty plant-based packaging to help cure the city's addiction to single-use plastic.

If you never got off the train, you'd be on board for a week. It is, however, perfectly possible to make stopovers – at Perm and Irkutsk, say – by buying separate tickets for each leg.

The two-night sailing from Vladivostok is aboard the *Eastern Dream*, which carries up to 500 passengers and 66 vehicles. With a jauntily retro nightclub on board, complete with high-energy dance tunes and coloured lights, it can be something of a party boat, but there's plenty of sleeping space, too, in private en-suite cabins or large dormitories with futons or bunks.

Sakaiminato is under 1,000km (621 miles) from Vladivostok, but you'll notice a profound cultural shift the moment you arrive. For anyone with a fascination for all things Japanese, the real adventure begins here. If Tokyo is your main focus, you could snooze your way there on the 15-hour night bus. But there's much to be said for travelling by train, breaking your journey at ultra-modern Osaka and elegant, historic Kyoto (*pictured, right*), then whizzing into the capital by Shinkansen bullet train.

As a side trip, travellers can take a steam train along the historic Circum-Baikal Railway, once dubbed 'the golden buckle on the steel belt of Russia'.

ALSO TRY

Trans-Mongolian Railway: Leave the Trans-Siberian Railway at Ulan-Ude in Siberia and head south, following the old tea-caravan route to Ulaanbaatar, the Gobi Desert and Beijing.

Trans-Manchurian Railway: Change at Chita for an alternative route to Beijing. It travels south via Harbin, home to the world's biggest winter festival, with ice sculptures galore.

Donghae and Seoul, South Korea: The cruise-ferry from Vladivostok to Sakaiminato sails via Donghae. Disembark here, and explore: it's less than three hours from South Korea's vibrant capital by train or bus.

The ultimate overland adventure: experience Africa from top to toe

EUROPE AND AFRICA

A trip from Amsterdam to Cape Town guarantees remarkable encounters and an unbeatable sense of achievement

Most of the planes that fly south from Europe to Africa take off in the evening and travel in darkness. Their passengers arrive in the early hours: excited, perhaps, but groggy from a broken night's sleep. Tremendous landscapes – the shifting dunes of the Sahara, the chattering forests of the Congo Basin – slipped away beneath them, unseen, during the flight. Never has a whole continent's teeming, confounding interior been so overlooked.

By contrast, an overland trip reveals Africa in stark, intimate close-up. For many of those who attempt it, this is a once-in-a-lifetime journey, full of vivid encounters and experiences that would be totally out of reach in other circumstances. For others, the thrill of the continent's pungent markets, bustling towns, starry skies and wildlife-rich wilderness is addictive, leading to years of diligent saving, ready for the next chance to return – even if just for a brief safari holiday.

Overlanding across Africa is not to be taken lightly, however. Africa's rail network is extremely fragmented, so travelling by road is the only practical option. The uninitiated could easily underestimate the continent's vast scale and the many challenges it can present, from red tape, health scares, unexpected expense, civil unrest and dangerous animals, to culture shock. That said, plenty of first-time overlanders who leave Europe armed with minimal knowledge and experience end up arriving safely in South Africa, flushed with success. Some feel so at home on the road that, a few Castle Lagers later, they're plotting an overland journey back.

Making a plan

Travel extensively in Africa, and you'll regularly hear the phrase 'we'll make a plan'. It's a favourite response to unexpected challenges. To make your trip as enjoyable as possible, it's wise to channel this can-do spirit right from the start.

How will you travel? On major routes, public transport by shared taxis, minibuses and buses tends to be abundant but gruelling, especially where roads are in poor condition. The greenest alternatives, walking or cycling, are not impossible, assuming

THE LOWDOWN

Best time of year:
For the easiest road conditions, avoid the West African rainy season (June to October) by departing between September and December. High season in Cape Town is November to February. The best time for a wildlife safari in southern Africa is the dry season, April to September.

Plan your trip: Allowing at least six months for the journey gives you time to explore as you travel.

Getting there:
Amsterdam is around 420km (260 miles) by road and sea from London, via the ferry from Harwich to Hoek van Holland. Alternative vehicle crossings from the UK with links to Amsterdam include the Eurotunnel motorail from Folkestone and ferries from Dover, Hull and Newcastle. A flight-free return trip from Cape Town to Europe, travelling non-stop as a passenger on a cargo ship, would take around 16 nights.

Deep in the heart of Africa, Uganda's Bwindi Impenetrable Forest is tough to reach overland and challenging to explore on foot, but the rewards are great: it's home to mountain gorillas.

Organized trips across Africa are great fun for those who enjoy mucking in. They typically depart from London and take around 23 weeks to reach the Cape. There'll always be something to do, be it market shopping, striking camp or digging the vehicle out of the mud. Going it alone, with or without guides, means missing out on this kind of camaraderie, but the rewards of tailoring everything to your personal interests may more than compensate.

you can spare the time. Travelling by electric vehicle is viable, too: you'll need one that's rugged and adapted for the purpose, with extra batteries. Even travelling by diesel-powered 4WD or motorbike can be greener than flying: the secret is to journey mindfully, rather than simply charging through. Visit remote ecotourism projects, and you can help African communities preserve fragile wilderness regions that capture carbon and support biodiversity.

Whatever you choose, you'll need to put in some careful research to ensure your route is safe, your jabs, anti-malarials and paperwork (including insurance and visas) are in order and you have all the camping, medical and mechanical equipment you'll require.

First steps to the Cape

It was in 1652 that Dutch traders founded Africa's first European settlement on a patch of Khoi San territory at the foot of Table Mountain. The traders' vegetable plots and orchards have remained green ever since and are now a city park, Company's Garden, named after the Dutch East India Company.

Opposite: Bloubergstrand Beach, north of Cape Town, South Africa, has superb views of Table Mountain. Below: uniquely adapted to survive the arid conditions of the Namib Desert, oryx are a common sight in Sossusvlei, Namibia.

WATER-WISE
CAPE TOWN

The Western Cape has recovered from the drought that struck in the summer of 2017–8, but water-saving habits have stuck, and it's good practice to follow suit. At attractions such as Cape Town's superb Zeitz Museum of Contemporary Art Africa (Zeitz MOCAA), notices remind you to use hand sanitizer rather than water to wash your hands.

MAKE A DIFFERENCE

You're sure to encounter people and wildlife projects in need during your trip. Resist the temptation to hand out sugary sweets to wide-eyed kids, and spend some time with them instead: a kick-about with a ball you've brought with you or a friendly chat to help them practise their English will go down far better with their parents. If you'd like to donate to a good cause, try to seek advice from local charities, schools or conservationists. The African people and programmes that would benefit most from visitors' support don't necessarily approach strangers with their hands outstretched, literally or metaphorically.

The tumbling Epupa Falls on the Kunene River, marking the border between Angola and Namibia, are at their most impressive in April and May.

Less than 14 degrees of longitude separate Amsterdam and Cape Town, but they're almost a hemisphere apart: a distance of over 86 degrees latitude, spanning 7,359 nautical miles via the Atlantic Ocean or around 13,000km (8,000 miles) straight across the Sahara. The safest and most enjoyable route for present-day over-landers is closer to 16,000km (9,900 miles), looping along the West African coast before crossing the equator.

Tempting as it may be to linger on your way across Europe, the pull of Africa is sure to be strong. One of the easiest entry points is Tangier in Morocco, reached via the vehicle ferry across the Strait of Gibraltar from Tarifa in southern Spain. It's as you approach the dock, and Africa comes into focus for the first time, that the adventure really begins.

Africa unfolds

A typical route from Tangier to Cape Town devotes around four weeks to the medinas and deserts of Morocco and Mauritania, seven weeks to the vibrant and supremely musical nations of sub-Saharan West Africa, six weeks to central Africa's humid forests and truck-eating potholes, and five weeks to austerely beautiful southwest Africa. On some days, you'll barely see a soul; on others, you'll be immersed in local life. Africa's climate encourages people to spend time outdoors, making communities extremely accessible.

Every traveller discovers their own highlights. They could be the evenings you spend dancing to mbalax music in Senegal's sweaty, racy nightclubs, or the mornings you rise early to watch bee eaters and herons on the mangrove creeks fringing the River Gambia. Maybe you'll love searching for chimpanzees in the forests of Guinea, or perusing traditional textiles in Ghana's busy markets.

Travelling through little-visited Cameroon, Gabon, Republic of Congo and Angola, you could hike through rainforests where elephants and lowland gorillas browse in the shadows, then visit Luanda, for a taste of urban beach life, well off the beaten track. Namibia is a country that merits revisiting again and again – its time-sculpted dunes, shimmering haze and uniquely adapted desert species are endlessly intriguing – and the spacious beaches of South Africa's Western Cape Province are the perfect prelude to breezy, beautiful Cape Town.

ALSO TRY

South Africa: With a vibrant food and cultural scene, beautiful beaches and superb wildlife-watching opportunities – from whales, sharks and penguins to lions, elephants and wild dogs – there's enough here to keep you busy for several weeks. If you didn't arrive in your own vehicle, you could easily rent one. For a treat, splash out on a two-day trip on the Blue Train.

Self-drive safari, southern Africa: Keep the momentum of your trip going by exploring South Africa's neighbours, Namibia, Botswana, Zimbabwe and Mozambique, visiting community-owned conservancies for superb wildlife-watching that benefits marginalized rural people.

Cape Town to Nairobi overland: Drive north via the safari heartlands of southern and East Africa, stringing together magnificently diverse destinations such as Victoria Falls, Lake Malawi, Zanzibar, the Serengeti and the Bwindi Impenetrable Forest.

Harness the elements on a transatlantic yacht rally

ATLANTIC OCEAN: USA TO PORTUGAL

Test your sailing skills on a long-distance voyage, meeting up with other crews at ports along the way

Only a tiny minority of amateur sailors get the opportunity to race around the globe. But a leg of the Atlantic Rally for Cruisers (ARC) is perhaps the next best thing. It offers the safety and companionship of sailing in a large fleet along a challenging but manageable offshore course.

The fortunate few who have first-class sailing, provisioning and navigational skills – and 15 months to spare – can join the World ARC, a 48,000-km (26,000-nautical-mile) global circumnavigation held annually since 1986. The complete course rides the trade winds from St Lucia via the Panama Canal, the Pacific, Australia, the Indian Ocean, Cape Town, St Helena and Salvador back to St Lucia. It's also possible to tackle just half the course, or pick one of half a dozen or so shorter annual ARC rallies.

One such is ARC Europe, a convivial event, sailing from Portsmouth on Chesapeake Bay in Virginia (or, if you prefer, from Nanny Cay on Tortola in the British Virgin Islands) to Bermuda, then crossing the Atlantic to the Azores and southern Portugal. While a competent crew could easily tackle this on its own, there's much to be said for an organized crossing.

Participants can keep in touch by sharing daily logs and, at every port, there's a team of World Cruising Club 'yellow shirts' on hand to offer advice and technical support. These helpers also organize parties, sightseeing tours and awards for speed and seamanship.

For some skippers, speed is of the essence, but the rally is open to less competitive sailors, too, and the itinerary allows ample time to get to know Bermuda and the Azores.

THE LOWDOWN

Best time of year: For favourable conditions when sailing from west to east across the Atlantic, depart the USA or Caribbean in April or May. ARC Europe, the annual rally to Portugal, departs in May. When making a westbound crossing, depart in December or January. Avoid hurricane season, which is June to November.

Plan your trip: A one-way journey takes around four weeks direct on a 10–30m (30–90ft) yacht, or seven weeks including stops.

Getting there: For a flight-free round trip from the UK by yacht, prepare for the rally by making a westbound Atlantic crossing around the turn of the year, spending the remainder of the winter and early spring in the Caribbean or on the US east coast.

Hidden gems of the North Atlantic

Set apart from the islands of the Caribbean, Bermuda has a distinctive culture and character. Its beaches and rum cocktails are rightly famous, and the air in the botanical gardens is sweet with roses and frangipani. Jump on a bike in St George's, the former capital, to explore cobbled streets lined with British colonial buildings, seeking out dishes such as spiny lobster and fish chowder in local restaurants.

The small town of Horta, on Faial in the Azores, is one of the busiest sailors' stopovers in the world. The walls and pavements of its marina are covered in colourful murals, painted by mariners to bring them good luck. You'll have time to visit other parts of this underrated archipelago, too, sampling wine made from grapes grown in Pico's volcanic soil, and enjoying gardens, thermal springs and lanes clogged with hydrangeas on São Miguel (*pictured, right*). The route ends on a celebratory note at Marina de Lagos (*pictured, below*), gateway to Portugal's Algarve.

ALSO TRY

ARC Portugal: This sailing rally covers 1,000km (550 nautical miles) from Plymouth in England across the Bay of Biscay to Baiona in Spain, then port-hops another 640km (345 nautical miles) south to Lagos in Portugal.

ARC Baltic: Covering 3,330km (1,800 nautical miles), this group sail visits ports and archipelagos in Germany, Denmark, Estonia, Russia, Finland and Sweden, to experience both the natural beauty and the cultural history of the region.

Board a tall ship for a low-carbon cruise on the high seas

ATLANTIC OCEAN: CANARY ISLANDS TO THE CARIBBEAN

Feel the salty breeze in your hair as you cruise the Atlantic, watching for whales and dolphins in the waves

It's hard to imagine a more romantic way to travel between the sun-soaked Canary Islands and the Caribbean than by tall ship, its sails billowing in the stiff Atlantic breeze.

The voyages on offer alter from time to time, both in the ports they connect and in which ships are plying the routes. One example is the *Star Flyer*, a modern, 100-m (360-ft) clipper designed to evoke the golden age of long-distance sailing, which takes you from the port of Las Palmas de Gran Canaria to Philipsburg on the Dutch island of Sint Maarten. After departing Las Palmas, the ship calls at San Sebastián on La Gomera, then spends 12 days at sea on its journey west, travelling under sail whenever conditions allow (which, on this classic mariners' route, is most of the time).

The *Star Flyer* is typical of the passenger tall ships that cruise to and from the Canary Islands. With four masts, 16 sails, teak decks and mahogany rails, it's sleek, elegant and graceful. Below deck, the fittings have a nostalgic, vintage feel, with varnished wood, polished brass, antique prints and paintings of famous ships. As well as being significantly more eco-friendly than a fuel-powered cruise ship, the *Star Flyer* is more intimate; rather than accommodating three thousand or so passengers, it has cabins for just 170. This difference in scale gives tall ship sailing an exclusive feel, and this is reflected in the price, which is around forty per cent more than a conventional cruise.

You're certainly well looked after. There are quarters for a 70-strong crew. You're not expected to help them weigh the anchor or hoist the mizzen, but they're happy to chat about the art of tall ship sailing when you meet them on deck.

THE LOWDOWN

Best time of year: Eastbound voyages depart from the Canary Islands in November and December. Voyages following a similar route in reverse depart from the Caribbean in March.

Plan your trip: The Atlantic crossing takes around 15 days. To enjoy the islands, allow at least four days in the Canaries and four in the Caribbean. Ferries connect Sint Maarten to Anguilla, St Barts and Saba.

Getting there: Gran Canaria, west of Morocco, is around 3,600km (2,200 miles) by road and sea from London. By train, travel across France and Spain to Cádiz or Huelva for a ferry to Las Palmas (four departures weekly, 32–36hr). Allow at least three nights: in Barcelona, Seville and on the ferry, for example. Options for a flight-free return journey include a berth on a transatlantic cargo ship or yacht, or chartering a boat.

The rugged volcanic landscapes of La Palma in the Canary Islands lend themselves to hiking.

HIKE THE CANARIES

A challenging new 651-km (404-mile) hiking trail, the GR131, allows you to explore all seven of the Canary Islands on foot, island-hopping by ferry. You could choose a section that appeals or, with five weeks to spare, cover the entire archipelago.

IN THE KNOW

For a peaceful experience, steer clear of Sint Maarten's Maho Beach: it's so close to the end of the airport runway that sunbathers are blasted by the jet engines of aircraft taking off and landing. Head for beautiful Cupecoy Beach or Mullet Bay (*pictured, right*) instead.

Long, lazy days at sea

Some first-time passengers wonder how on earth they will fill their days, but their doubts soon evaporate. Your ship is sure to be comfortable; surprisingly, perhaps, the *Star Flyer* has two swimming pools. As on any cruise, there's also a certain amount of organized entertainment, from lectures and quizzes to deck parties and games, with extra events popping up depending on the passengers' inclinations.

While you're close to the Canary Islands, keep your binoculars at the ready: around one-third of the world's whale and dolphin species are found here. Later, if you're blessed with fine weather – which is always a strong possibility – you can simply relax on deck, relishing the absence of engine noise.

To really bond with your ship, you could stretch out on the bowsprit nets or, safely strapped in a harness, climb the mast to the crow's nest to ponder the ocean's enormity. Come nightfall, there's another treat in store: the North Atlantic has famously clear skies, so don't miss the chance to gaze up at the stars, just as the great navigators used to do.

ALSO TRY

Bremerhaven to Hendaye: Voyage by tall ship from Bremerhaven in northern Germany – a shipbuilding city that hosts major maritime festivals – to the Basque resort of Hendaye on the Bay of Biscay in southwest France.

Oslo to Lofoten: Take part in a long-distance tall ship race along Norway's crinkled coast, joining the international crews for a few days as they battle it out in mini regattas.

Singapore to Bali: Join a tall ship expedition across the Java Sea, visiting some of the 17,000 islands in Indonesia, the world's largest archipelago.

Voyage from Italy to Australia by cargo ship

ITALY, RED SEA, INDIAN OCEAN AND AUSTRALIA

Retrace one of history's major migration routes, through the Suez Canal and across the Indian Ocean to the southern hemisphere

Looking for the best espresso bar in Australia? Of the many baristas who would fight tooth and nail for that title, some of the top contenders are undoubtedly in Melbourne's Little Italy. It's home to a good number of the million-strong Italian Australian community, whose great-grandparents lived 16,000 km (9,950 miles) away. They've dispersed a little in recent years, but on Lygon Street, plenty of Italian delis, cafés and restaurants remain, proudly decked out in red, white and green. Ever since Australia's first pizzeria, Toto's, opened on Lygon Street in 1961, there's also been fiery competition for the best Aussie pizza: chunkier than its Italian counterpart, with alternative toppings such as bacon and eggs.

Italian economic migrants arrived in large numbers between the 1850s and the 1960s, drawn by gold mines, sugar cane fields and other opportunities. For post-war migrants, the passage cost the equivalent of an executive's annual wage, forcing families to endure long separations as individuals made the move one at a time. For some, the fare and basic accommodation were covered by the Australian government, but they had to put in several years' work in return.

The earliest ocean voyages from Genoa, Naples or Messina via the Cape of Good Hope must have seemed endless. Even in the mid-20th century, decades after the Suez Canal had opened, the passage took six to eight weeks. Modern freighters have shaved that time down, however, and travellers can now book a cargo ship voyage lasting around five weeks.

THE LOWDOWN

Best time of year: Depart between December and April to avoid the Indian Ocean monsoon (May to November). High season in Australia is December to February.

Plan your trip: A one-way voyage to Fremantle, Adelaide, Melbourne, Sydney or Brisbane takes five to seven weeks. Itineraries and shore time (for which you may need to arrange visas in advance) depend on the weather and other variables.

Getting there: Italy's major container ports, including Genoa, Gioia Tauro, Naples and Palermo, are on major road and rail routes. From the UK, trains from London St Pancras via Paris connect with routes throughout Italy, including the sleeper train from Milan and Rome to Palermo in Sicily. As an alternative to flying back from Australia, you could return via the Far East, or continue across the Pacific to the Americas as part of a flight-free round-the-world trip.

WHAT ABOUT THE IMPACT?

Cargo ship passengers argue that, since its primary purpose is to transport freight, their ship would travel whether they were on board or not. Their voyage is therefore effectively carbon neutral. Is that greenwash? Hopefully, in the future, this question will evaporate. Shipping companies are seeking alternatives to heavy fuel oil, a fossil fuel that pollutes the air with high levels of sulphur oxide and nitrogen oxide emissions. China has already launched its first electric freighters, and concepts in development include eco-ships powered by solar cells, wind, waves, liquefied natural gas and hydrogen fuel.

Life at sea

A journey as long as this remains the essence of slow travel. There are many days at sea, so it helps to be self-sufficient. Facilities may be limited to a lounge and dining room, a basic gym and perhaps a small swimming pool. On some ships it might be possible to tour the engine room and bridge, at the discretion of the captain and chief engineer. Overall, your voyage could be the perfect time to get stuck into some serious reading, writing or study.

There will, however, be a few stops to break up the journey, and you may have a full day on shore at each, again at the captain's discretion. Boarding your ship at a busy commercial port such as Genoa in northern Italy or Gioia Tauro in the south, you'll sail to Port Said or Damietta in Egypt; if your ship docks here for a full day, there would be just enough time to take a taxi to the spectacular new Grand Egyptian Museum outside Cairo.

Opposite: Genoa, home to the Mediterranean's biggest and busiest port, has been a seafaring city for centuries. Below: The soaring minarets of Port Fouad's waterfront mosque mark the entrance to the Suez Canal in Egypt.

BAG YOUR BERTH

Several specialist agents offer a cargo ship cruise booking service. They all apply the same tariffs (currently around €150 per person per day, full board), and special offers are rare. Freighters plying the routes between Europe and Australia may have as few as three two-person passenger cabins available, so it's crucial to make enquiries as far in advance as you can.

You'll then sail through the Suez Canal down the Red Sea, perhaps stopping at Djibouti or Oman. Next, you'll cross the Indian Ocean, possibly via Sri Lanka and Singapore, with a chance to visit the Sri Lankan capital Colombo and Singapore's Gardens by the Bay. An alternative route takes you via La Réunion and Mauritius, for a brief glimpse of Indian Ocean island life. When, after the long days of open ocean, you finally reach Australia, ports of call may include Fremantle, Adelaide, Melbourne and Sydney.

A taste of Italy in Australia

Melbourne isn't the only Australian city with an Italian population: Sydney (*pictured, left*), Brisbane and Adelaide all have vibrant communities. Although many of the first migrants came from Sicily and Calabria in the far south, today, most Italian Australians trace their ancestry to the northern Italian regions of Veneto, Piedmont, Lombardy and Tuscany: many migrants came from these areas to work in the Australian mines and plantations.

The importance of espresso to Italian Australian culture bears repeating. Italians seeded Australia's obsession with good coffee, and 1950s espresso bars such as Pellegrini's in Melbourne and Piccolo in Sydney still hold their own against the tide of hipster newcomers. Just don't expect anyone to agree on exactly which one is best.

ALSO TRY

To Australasia via China: Make your way to Moscow and ride the Trans-Siberian Railway (see page 204) across Russia, changing onto either the Trans-Mongolian or Trans-Manchurian lines for Beijing. Next, hop on the world's fastest mainline train, the Fuxing Hao, to zoom to Shanghai in 4hr 18min. From Shanghai, cargo ships sail to ports in Australia and New Zealand including Melbourne, Sydney, Brisbane, Auckland, Napier and Port Chalmers.

To Australasia via Japan: Take the Trans-Siberian Railway across Russia to Vladivostok, the ferry to Sakaiminato (Japan), and the train to Yokohama, then board a cargo ship across the Pacific Ocean.

To Australasia from the USA: Board a Pacific cargo ship in Philadelphia or Charleston to travel via Cartagena (Colombia) and the Panama Canal. Alternatively, start from Oakland or Los Angeles in California.

Head north on an expedition to Iceland and Greenland

GERMANY, NORWAY, ICELAND AND GREENLAND

Embark on a climate-conscious cruise to the Arctic Circle to discover remote coastal landscapes, lit by the midnight sun

What is it that drives us to achieve the near-impossible? The Norwegian explorer Fridtjof Nansen began hatching the idea of crossing the interior of Greenland in his early twenties, while studying Arctic wildlife in the Greenland Sea west of Spitsbergen. Since he was an expert cross-country skier, it was perhaps not too far-fetched an ambition. But nobody had completed such an expedition before.

Today, Greenland has a population of barely 56,000, most of them Inuits, who live in the capital, Nuuk, or in coastal villages. It remains one of the few countries on Earth that's bitterly challenging to cross, and in the absence of international ferries, it's extremely difficult to reach without flying. Difficult, but not impossible. Tourists can explore its beautiful settlements and bays by joining an expedition cruise on an eco-friendly 530-passenger ship that bears Nansen's name.

Hurtigruten's MS *Fridtjof Nansen* was launched in 2020. It's one of the world's first hybrid cruise ships, powered by batteries and low-sulphur marine gas oil. Thanks to its high-tech engines, streamlined icebreaker hull and energy-efficient heating, cooling and water-management systems, it uses 20 per cent less fuel than conventional cruise ships of a similar size, with a corresponding reduction in greenhouse gas emissions. As a bonus, it's quiet, particularly when running on batteries, with reduced drag, noise and vibration.

Making cruising cleaner and greener once seemed impossible, but Hurtigruten has proved that the industry can change for the better. Its ultimate goal is to launch a fleet of zero-emissions ships.

THE LOWDOWN

Best time of year: Expedition ships cover this route, and its reverse, in the northern summer (June to September).

Plan your trip: The journey combines two ocean expeditions: a 15-day cruise from Hamburg to Reykjavík, and a 16-day trip from Reykjavík to Greenland and back.

Getting there: Hamburg is around 930km (580 miles) by road and sea from London via Belgium. High-speed trains from London St Pancras via Brussels take around 9hr 30min. Ferries from Dover, Harwich, Hull and Newcastle to France, Belgium and the Netherlands are connected to Germany by road and rail. To travel from Reykjavík to the UK without flying, cross Iceland by bus to Seyðisfjörður (11hr 20min) for the ferry via the Faroe Islands to Hirtshals in Denmark (37hr), then continue by train and bus (23hr).

In midsummer, the sun never sets over Oqaatsut, a tiny village on Disko Bay, western Greenland.

FRIDTJOF NANSEN

The Norwegian explorer
and Nobel Peace Prize
laureate Fridtjof Nansen
led the team that made
the first crossing of
Greenland's icy interior.
Setting out in the summer
of 1888, they planned to
ski from Sermilik to Disko
Bay, but bad weather
prevailed. They switched
course, and made it from
Umivik to Godthaab
(now Nuut), a journey of
over 400 km (250
miles), in 49 days,
battling through fresh
snow and temperatures
that plummeted to
-45°C (-49°F) at night.

SHIPPING FUELS
OF THE FUTURE

Heavy fuel oil, the
cheap, polluting fossil
fuel that some cargo
and cruise ships still use,
was banned in the
Antarctic in 2011. A
similar ban will apply in
the Arctic from 2024.
Cleaner alternatives
include marine gas oil,
which is widely
available, and liquefied
natural gas. Renewables
such as hydrogenated
vegetable oil and liquid
biogas, which can be
produced from used
cooking oil or organic
waste from the Nordic
fishing and timber
industries, reduce harmful
emissions by 90 per cent.

An eco-cruise in subarctic waters

The MS *Fridtjof Nansen*'s interior sets the tone for the trip. Inspired by Scandinavian landscapes, it features natural materials such as granite, oak, birch and wool. The restaurants offer Nordic menus designed to minimize waste, and there's no single-use plastic anywhere on board. Guides give lectures on cultural history, ecology and the observable consequences of climate change, and there are observation decks for wildlife-watching. But you won't just be watching the scenery roll by: you'll be exploring by small boat, kayak or on foot.

The journey takes you into majestic Norwegian fjords, quaint Arctic villages and wildlife-rich seas. From Hamburg, one of Europe's greenest cities, you'll cruise to pretty Kristiansand in Norway, wandering the town or going whitewater rafting in the Setesdal valley. Further north, the scenery becomes more and more monumental, and you'll sail into Lysefjord, where the sheer 604-m- (1,980-ft)-high high cliffs of Pulpit Rock tower above. Next comes Geirangerfjord, whose deep blue waters, fairy-tale mountains and waterfalls inspired Disney's *Frozen*.

Crossing the Norwegian Sea to Iceland takes two nights. Some parts of this Nordic country, such as the Golden Circle near Reykjavík, have been afflicted with mass tourism in recent years; visiting by expedition ship allows you to explore quieter towns and wilderness regions instead, taking care to minimize your impact.

The first stop is Seyðisfjörður, where you could watch birds in Skálanes Nature Reserve or relax in Iceland's purest hot springs. Near Isafjörður and Patreksfjörður, you can tour the eco-progressive fishing village of Suðureyri and feel the spray of the Dynjandi waterfall. Stykkishólmur, the first European community to receive EarthCheck's gold certification for sustainable tourism, has a striking church and exciting hiking trails, and the island of Heimaey is a haven for puffins. You then double back to Reykjavík, ready for the two-day journey across the Denmark Strait to Greenland.

REYKJAVÍK

Powered by geothermal energy, the world's northernmost capital is one of the world's most sustainable cities. Don't miss the chance to take a dip in the North Atlantic followed by a relaxing soak in a thermal pool, letting natural algae and minerals benefit your skin.

Above: with a mild climate, the village of Narsarsuaq in southern Greenland has the territory's only botanical garden, the Greenlandic Arboretum. Opposite: icebergs created by the calving of the Jakobshavn Glacier in western Greenland have been the subject of a long-term study by glaciologists and climate scientists.

Along Greenland's wild west coast

You'll land at Qaqortoq, where you could, as the guest of a local family, learn about local lifestyles over a *kaffemik* – an open-house feast of homemade delicacies. In Kvanefjord, it's time to take to the water, kayaking among icebergs, looking for seabirds and seals. You'll cross the Arctic Circle to the traditional town of Sisimiut, climbing Palasip Qaqqaa mountain for views of the colourful houses below, then marvel at Ilulissat's remarkable ice-scapes. There are more glaciers, bergs and floes to admire at the route's northernmost point, Disko Bay.

Heading south again, the ship explores the dramatic landscapes of Evighed Fjord, then docks in Nuuk for a catch-up on cultural history. You'll visit the long-abandoned Norse settlement of Ivittuut – now ruled by shaggy-coated musk oxen – before returning to Reykjavík.

ALSO TRY

Reykjavík, Iceland: Despite a recurring surfeit of visitors, Iceland's capital has oodles of cool. Extend your stay to enjoy the capital's striking modern architecture and fine restaurants.

Lofoten and Tromsø, Norway: Enjoy a half-day cruise in Arctic waters in a hybrid boat, watching humpback whales, orcas, eagles or the northern lights.

Alaska's southern islands, USA: The Alaska Marine Highway ferry system allows you to explore.

Go with the floe: Barcelona to Antarctica

ATLANTIC OCEAN, SOUTH AMERICA AND ANTARCTICA

THE LOWDOWN

Best time of year: Depart between October and February. Antarctic expedition cruises operate in summer, November and March.

Plan your trip: Allow at least six to eight weeks, including three to four weeks crossing the Atlantic and ten to fifteen days on an expedition cruise. Cargo ship voyages need to be booked well in advance. Your ship may not depart on schedule, so it's important to set aside plenty of contingency time.

Getting there: Barcelona is around 1,500km (930 miles) by road and sea from London via Kent and France. High-speed trains from London St Pancras via Paris take around ten hours. Ferries from the UK to France, Belgium and the Netherlands are connected to Spain by road and rail. For a flight-free return journey from Argentina to the UK, you could retrace your outward route, or travel overland to Brazil and cross the Atlantic from one of its cargo ports.

Follow in the wake of the great polar explorers on an epic southbound ocean voyage

Like the continent's adorable avian residents, penguins, Antarctic tourists don't fly. It's impossible to reach the mainland of the southern continent by plane, unless you're one of the 4,000 or so international staff who work at its scientific research stations: tourists travel by sea.

Admittedly, flying can get you pretty close. Some visitors travel by plane to the commercial airstrip on King George Island, the largest of the South Shetland Islands, just 120km (75 miles) north of the Antarctic Peninsula's rocky tip. They then board a ship to explore Antarctica proper. But they're missing out on a key part of the journey – the part that makes your brief visit begin to feel like an expedition – the Drake Passage between Cape Horn and the Antarctic Peninsula. Crossing it by sea is a must for any true adventurer who's approaching from South America.

This stormy stretch of water is 814km (440 nautical miles) wide and well over 3km (1.8 miles) deep. In a small expedition ship, the voyage from Ushuaia in Argentina or Punta Arenas in Chile, where most Antarctic cruises begin, takes a couple of days and nights – enough time to attend lectures on Antarctica's ecology, learn about all the protocols that help protect its fragile habitats, and check all your gear for possible contaminants, from grass seeds to grains of sand. It's also enough time, when conditions are challenging, for the ocean to thoroughly rearrange every item on the ship that isn't bolted down. When at last you arrive, you'll appreciate the calm splendour of the peninsula's pristine bays all the more.

Backed by the mountains of Tierra del Fuego, Ushuaia is the starting point for a sea voyage from Argentina to Antarctica via the Beagle Channel and the Drake Passage.

Argentina's capital city has beguiling, European-style boulevards and a buzzing, creative atmosphere. Many of the clichés are true: tango dancers really do busk in the colourful streets of La Boca (*pictured, right*), the smoky aroma of *parrillas* (barbecues) fills the air on warm evenings and there's always a crowd at Eva Peron's tomb in Recoleta. To help keep the public fit, pop-up Estaciones Saludables (health stations) in parks and squares offer free check-ups and exercise classes.

Crossing the Atlantic by cargo ship

But what about the journey that precedes this, to the gateway towns of Punta Arenas or Ushuaia? Flying is the norm, but travelling by ocean and land gives a far richer appreciation of Antarctica's geological connection to South America.

Starting from western Europe, Buenos Aires is an obvious first target. Cargo ships travel direct to South America from Barcelona and other major Mediterranean ports such as Valencia, Genoa and Gioia Tauro. On a typical itinerary, stopping at Salvador, Santos, Itapoá and Paranaguá, the journey to Argentina takes just over three weeks. It's an expensive way to cover the distance, but it offers the luxury of time at sea and the chance to glimpse gritty parts of Brazil you might otherwise never visit.

The 3,000-km (1,800-mile) RN3 highway from Buenos Aires to Ushuaia is a classic: it's the southernmost leg of the legendary Pan-American Highway from Alaska to Tierra del Fuego. You could drive it on two wheels or four, or take buses, changing at Río Gallegos for a connection to Ushuaia or Punta Arenas.

THE SCENIC ROUTE TO TIERRA DEL FUEGO

Rather than heading due south from Buenos Aires, you could travel west to Santiago de Chile then south to Puerto Montt, to take a cargo ferry to Puerto Natales. As you cruise the strikingly picturesque, little-visited channels of the Chilean coast, you might spot humpback whales, fur seals and Magellanic penguins: an excellent way to whet your appetite for Antarctica.

ANTARCTIC CRUISING

Cruise ships are controversial, but Antarctic ships take meticulous care to abide by the International Association of Antarctica Tour Operators' stringent guidelines on minimizing pollution, safeguarding wildlife and protecting the environment. There's also a strong push towards reducing emissions. The number of tourists allowed at each landing point is strictly limited, so the smallest ships offer the most excursions and landings. Larger ships may simply provide distant views from the deck.

Exploring the Antarctic Peninsula

Some visit Antarctica for the wildlife, or to learn about the great explorers. For others, it's the magnificent mountains and ice-scapes. But for many, it's simply a precious opportunity to visit a pristine location that's unlike anywhere else on the planet. Even the Arctic only bears a superficial resemblance. Beyond the obvious differences between their wildlife – where the Arctic has polar bears, foxes and walruses, the Antarctic has elephant seals, leopard seals and penguins – Antarctica has layer upon layer of distinctive natural history and scientific heritage.

Expedition ships visit a new part of the peninsula each day, working around the weather to ensure you have as much time as possible to explore – hopping into Zodiac boats to cruise up to monumental glaciers or around ice-scattered bays where whales breach in the distance and seals loll like mermaids on icebergs. The greatest privilege of all is to land on the peninsula or its islands, scrambling out of your Zodiac onto gravel beaches to tiptoe past slumbering elephant seals and cacophonous, pungent penguin colonies. Antarctic animals are generally unfazed by human visitors, as long as you don't approach closer than 5m (16ft) or so. This is without a doubt one of the most extraordinary wildlife-watching experiences the natural world can offer.

ALSO TRY

South Georgia and Falkland Islands: Extend your time in the southern seas with an expedition to these fascinating, far-flung islands, to clock up more sightings of albatrosses, penguins and seals.

Alaska: Marvel at mighty glaciers on a wilderness voyage from British Columbia, with a chance to see bald eagles and brown and black bears.

Svalbard: Join an ocean expedition to Norway's High Arctic archipelago in a modern, eco-friendly ship, for polar bear sightings and the chance to make a positive difference by taking part in beach clean-ups.

Footloose and flight-free: circle the world by land and sea

ROUND THE WORLD

THE LOWDOWN

Best time of year:
Any time, depending on your chosen route and starting point. In both hemispheres, spring and autumn are usually best. It's wise to avoid hurricane and monsoon seasons when making long-distance journeys by sea.

Plan your trip: To make the most of the journey, allow at least eight to ten months. Even if you'd like to keep your itinerary fluid, fix critical aspects of your trip in advance, such as ocean crossings, rail passes, visas and vaccinations.

Getting there:
Commercial ports and railway hubs are likely to be key points on your itinerary. The busiest ports include Vancouver, Los Angeles, Rotterdam, Antwerp, Cape Town, Shanghai, Singapore and Port Hedland (Australia). Major stations include Los Angeles, Chicago, Montreal, New York, London, Paris, Hamburg, Vienna, Moscow, Beijing, Shanghai and Bangkok.

Meet people, enjoy nature and make spontaneous discoveries as you travel at a gentle pace, creating memories to last a lifetime

Tell someone that you're planning to travel round the world without flying and, chances are, their first question won't be: why? After all, there's been plenty of discussion of the downsides of flying in recent years, and most of us love a challenge. No, the first question will probably be: how? When you spin the globe, an enormous proportion of it is blue.

Unless you have the luxury of unlimited time and money, giving careful consideration to those all-important ocean crossings should be your first step. The finer details – what to pack, what insurance to buy and whether to create a video blog about the whole adventure – can come later.

If one of your aims is to harm the planet as little as possible, you should avoid conventional cruise ships, since they're among the worst polluters. Crossing oceans under sail or booking passenger berths on cargo ships are excellent alternatives. Both require advance planning: opportunities are limited and schedules can be irregular and unreliable, dependent as they are on tricky variables including the weather. Ocean crossings will also demand a large chunk of your budget, unless you're earning your passage on a yacht as crew. It's not generally possible to work on a cargo ship on a casual basis.

For some, circling the globe is a goal in itself. Others have an additional objective, such as visiting destinations with a common theme, or with cultural links to their home region. You could line up a series of volunteer placements, or use your adventure as a platform to raise money or awareness for a charitable cause. The following routes from Europe are just examples, to get you thinking.

GAME ON

On any long journey, there will be times when conversation falters and you're tired of the view. Bring plenty of music and a headphone splitter (so two can share), card games, writing and sketching materials, a well stocked e-book reader and a tennis ball or inflatable football – perfect for an impromptu kickabout with local kids.

Explore our green planet, one city at a time

What better way to kick off a tour of the world's greenest cities than by soaking up the Nordic nations' forward-thinking vibes? Hop between Copenhagen, Oslo, Stockholm and Helsinki by train and ferry, feasting your eyes on their futuristic architecture and gorging on sustainable food. Then take the train to Moscow and continue across Russia, China and Southeast Asia to Singapore, home to high-tech eco-hotels and the Gardens by the Bay.

A cargo ship will transport you from Asia to Adelaide, Australia's most eco-conscious city. From here, you could tour the East Coast's greenest enclaves, then take another ship to Japan to hang out in Tokyo (*pictured, right*). Cross the Pacific to the characterful, endlessly hip cities of Vancouver or San Francisco (*pictured, below*), then catch trains or long-distance buses to the Eastern Seaboard for an Atlantic crossing back to Europe.

STAYING OK ON THE ROAD

Mental health is a woefully under-discussed aspect of long-distance travel. If you find yourself struggling, meaningful conversations with friends and family over video or voice calls can be a lifeline, so don't scrimp on these. Maintaining a fitness regime is crucial, too. If culture shock, delays or indecision scupper your inbuilt need to feel in control of things, take time out to make lists or sketch out a few long-term plans. Most of all, seek positive connections with nature, wildlife and the people around you. Helping others can often cause your own problems to recede.

A world of wildlife-watching experiences

Start in the mountains of Poland or Romania, home to bison, lynx (*pictured, above*), wolves and bears. Then travel to a North Sea or Atlantic port for a voyage to Walvis Bay in Namibia or Cape Town in South Africa. Both are excellent launch points for an African wildlife safari. Continue across the Indian Ocean to Singapore, ferry-hop to Borneo and steel yourself for the long bus journey to Sabah: your reward will be quality time with orangutans in the wild.

After backtracking to Singapore, take another cargo ship to Australia or New Zealand to watch marsupials and Antipodean birds, then cross the Pacific to Colombia. From here, you could overland north to the bird-rich forests of Costa Rica, plunge south to the Peruvian Amazon, or both. For the ultimate wildlife adventure, take buses all the way south to Tierra del Fuego for an expedition cruise to Antarctica. Make for the Brazilian coast when you're ready to take a ship back to Europe.

Mosaic of cultures

Overland travel can immerse you in beautiful, remote regions, off the tourist trails. Even in busy western Europe, there are gems: the Basque Country and Brittany, for example, have delightful villages where you can make friends over a glass of wine.

From northern France or the Low Countries, take a cargo ship to Charleston, a charming city that tackles the complex history of the American Deep South. Tour the southern states by train, ending up in California, home to bohemian coastal communities. Then head north by road or rail to British Columbia, where an Indigenous Tourism network offers craft workshops and boat tours.

Container ships from Vancouver cross the Pacific to Shanghai: from here, you could spend weeks exploring the fascinating villages of China and Southeast Asia before returning to Europe by train.

THE GREEN TRAVELLER'S DIRECTORY

WORLDWIDE

HOLIDAYS & EXPERIENCES

ETHICAL TOURS & ADVENTURES
Audley Travel *audleytravel.com*;
Dragoman Overland *dragoman.com*; **The Explorations Company**
explorationscompany.com; **Explore**
explore.co.uk; **G Adventures**
gadventures.com; **Intrepid Travel**
intrepidtravel.com; **Much Better
Adventures** *muchbetteradventures.com*; **Original Travel** *originaltravel.co.uk*; **Responsible Travel**
responsibletravel.com; **Steppes
Travel** *steppestravel.com*

WALKING, HIKING & WINTER SPORTS
HF Holidays *hfholidays.co.uk*;
The Natural Adventure Company
thenaturaladventure.com; **Snow
Carbon** *snowcarbon.co.uk*; **Walks
Worldwide** *walksworldwide.com*

CYCLING & E-BIKING
Cycling for Softies *cycling-for-softies.co.uk*; **Saddle Skedaddle**
Cycling trips *skedaddle.com*; **The
Slow Cyclist** *theslowcyclist.co.uk*

RAIL
Belmond Venice Simplon-Orient-
Express and other hotel-trains
belmond.com; **Ffestiniog Travel**
Rail holidays *ffestiniogtravel.com*; **Golden Eagle Luxury Trains**
goldeneagleluxurytrains.com;
Great Rail Journeys Tours and
trips *greatrail.com*; **The Railway
Touring Company** Train holidays
railwaytouring.net

OCEAN, SEA & RIVER
Albatros Expeditions Antarctic
and Arctic *albatros-expeditions.com*; **AmaWaterways** River cruises
in hybrid ships *amawaterways.com*; **Aqua-Firma** Marine and
wilderness trips *aqua-firma.com;*
Fair Ferry Long-distance transfers
by sailing yacht *fairferry.com*;
Gulet Escapes *guletescapes.com*;
Hurtigruten Ocean cruises in hybrid
ships *hurtigruten.com*; **Slowtravel**
Marine adventures *langsamreisen.de*; **Star Clippers Cruises** Tall
ships *starclipperscruises.co.uk*;
Sunsail Yacht charters and sailing
sunsail.co.uk; **SwimTrek** Swimming
holidays *swimtrek.com*; **Viking River
Cruises** Holidays in hybrid ships
vikingrivercruises.com; **VoyageVert**
Ocean sailing *voyagevert.org*; **Wild
Earth Travel** Small-ship cruising
wildearth-travel.com; **Wildsea
Europe** Marine ecotourism *wildsea.eu*

WILDLIFE & NATURE
Blue Sky Wildlife *blueskywildlife.com*; **Natural World Safaris**
naturalworldsafaris.com; **Naturetrek**
naturetrek.co.uk; **Rainbow Tours**
rainbowtours.co.uk; **Wildlife
Worldwide** *wildlifeworldwide.com*

CULTURE
Eatwith Eating in private homes
eatwith.com; **Mamaz Social
Food** Culinary experiences
mamazsocialfood.com; **Martin
Randall Travel** Arts and gastronomy
martinrandall.com; **Undiscovered
Destinations** Culture and adventure
undiscovered-destinations.com

VOLUNTOURISM
Biosphere Expeditions Citizen
science *biosphereexpeditions.org*; **Earthwatch** Conservation
expeditions *earthwatch.org*; **GVI**
Study and research *gvi.co.uk*; **People
and Places** Ethical volunteering
travel-peopleandplaces.co.uk; **World
Wide Opportunities on Organic
Farms** *wwoof.org*

ACCOMMODATION
Bouteco Design-led eco-hotels
bouteco.co; **Canopy & Stars**
Glamping *canopyandstars.com*;
Ecobnb Rentals and experiences
ecobnb.com; **Green Pearls** Hotels
and restaurants *greenpearls.com*;
Homestay *homestay.com*; **Hostelling
International** *hihostels.com*; **My Eco
Stay** Holiday rentals *myecostay.eu*;
Nature House Rural rentals *nature.house*

TRANSPORT GUIDES
Bikemap Travel planning for cyclists
bikemap.net

Busbud Long-distance bus guide
busbud.com
Direct Ferries Ferry guide
directferries.com
Ecopassenger Emissions calculator
ecopassenger.org
Green Traveller Holiday planning
greentraveller.co.uk
The Man in Seat Sixty-One Train
guide *seat61.com*
Marine Traffic Ship tracker
marinetraffic.com
Naviki Navigation aid for cyclists
naviki.org
Open Railway Map Online route map
openrailwaymap.org
PlugShare Electric vehicle charging
points *plugshare.com*
Railcc Train guide *rail.cc*
Rome2Rio Travel planning *rome2rio.com*

CARGO SHIP TRAVEL AGENTS
Cargo Ship Voyages
cargoshipvoyages.com; **Freighter
Expeditions** *freighterexpeditions.com.au*; **Freightlink** *freightlink.co.uk*; **Maris** *freightercruises.com*;
Slowtravel *langsamreisen.de*

ROUTES & ORGANZATIONS
Abraham Path Walking route
through Turkey, Syria, Jordan,
Palestine & Israel *abrahampath.org*
BlaBlaCar Car-pooling network,
enabling drivers to share costs with
passengers *blablacar.com*
EarthCheck Sustainable tourism
certification *earthcheck.org*
European Cyclists' Federation
Cycling information, projects and
events *ecf.com*
European Ramblers' Association
European long-distance paths
(E-paths), GR routes, Leading
Quality Trails, Eurorando walking
festival *era-ewv-ferp.com*
EuroVelo Long-distance cycle routes
in Europe *eurovelo.com*
Via Francigena Walking route,
Canterbury to Rome *viefrancigene.org*
World Cruising Club ARC (Atlantic
Rally for Cruisers) yacht rallies
worldcruising.com

EUROPE & RUSSIA

TRANSPORT PASSES & AGENTS

Ferry Hopper Travel bookings *ferryhopper.com*
InterFlix Five-trip pass for FlixBus and FlixTrain *interflix.flixbus.com*
Interrail & Eurail Train passes and route-planning app (Interrail for Europeans, Eurail for non-Europeans). Global (33 European countries) or One Country (30 countries available) *interrail.eu, eurail.com*
Omio Travel bookings, comparing trains, buses, flights and ferries *omio.com*
Rail Europe Ticket and pass bookings *raileurope.com*

INTERNATIONAL BUSES

Alsa *alsa.com*; **BlaBlaBus** *blablabus.com*; **Busabout** *busabout.com*; **Eurolines** *eurolines.eu*; **FlixBus** *flixbus.com*

INTERNATIONAL TRAINS

ČD Czech Republic *cd.cz*; **CFR** Romania *cfrcalatori.ro*; **CP** Portugal *cp.pt*; **DB** Germany *bahn.com*; **Eurostar** UK *eurostar.com*; **Eurotunnel** UK *eurotunnel.com*; **HŽPP** Croatia *hzpp.hr*; **MÁV** Hungary *mavcsoport.hu*; **ÖBB** Austria *oebb.at*; **ÖBB Nightjet** Austria *nightjet.com*; **PKP** Poland *pkp.pl*; **RegioJet** Czech Republic *regiojet.com*; **Renfe** Spain *renfe.com*; **RZD** Russia *rzd.ru*; **SJ** Sweden *sj.se*; **Snälltåget** Sweden *snalltaget.se*; **SV** Serbia *srbvoz.rs*; **TCDD** Turkey *tcdd.gov.tr*; **TGV-Lyria** France/Switzerland *tgv-lyria.com*; **Thalys** Belgium/France *thalys.com*; **Thello** Italy *thello.com*; **UZ** Ukraine *uz.gov.ua*; **Urlaubs-Express** *urlaubs-express.de*; **ZSSK** *zssk.sk*

INTERNATIONAL FERRIES

Adria Ferries *adriaferries.com*; **Africa Morocco Link** *aml.ma*; **Akgünler Denizcilik** *akgunlerbilet.com*; **Anek Lines** *anek.gr*; **Aqua Bus** *aquabusferryboats.com*; **Baleària** *balearia.com*; **Blu Navy** *blunavytraghetti.com*; **Brittany Ferries** *brittany-ferries.com*; **Bulgaria Fast Ferry** *fastferry.bg*; **Color Line** *colorline.com*; **Condor Ferries** *condorferries.com*; **Corsica Ferries** *corsica-ferries.com*; **Corsica Linea** *corsicalinea.com*; **CTN** *ctn.com.tn*; **DFDS** *dfds.com*; **Dodecanese Flying Dolphins** *12fd.gr*; **Eckerö Line** *eckeroline.com*; **Erturk Lines** *erturk.com.tr*; **Finikas** *finikas-lines.com*; **Finnlines** *finnlines.com*; **Fjord Line** *fjordline.com*; **Fred Olsen Express** *fredolsen.es*; **FRS Iberia** *frs.es*; **Grandi Navi Veloci** *gnv.it*; **Grimaldi Lines** *grimaldi-lines.com*; **Inter Shipping** *intershipping.es*; **Irish Ferries** *irishferries.ie*; **Jadrolinija** *jadrolinija.hr*; **Liberty Lines** *libertylines.it*; **Manche Îles Express** *manche-iles.com*; **Mediterránea Pitiusa** *mediterraneapitiusa.com*; **La Méridionale** *lameridionale.com*; **Minoan Lines** *minoan.gr*; **Moby Lines** *mobylines.com*; **Moby SPL** *stpeterline.com*; **Navibulgar** *navbul.com*; **Naviera Armas** *navieraarmas.com*; **P&O Ferries** *poferries.com*; **PBM** *pbm.bg*; **Polferries** *polferries.com*; **Scandlines** *scandlines.com*; **Sea Dreams** *seadreams.gr*; **Sea Lines** *sealines.com.tr*; **Smyril Line** *smyrilline.com*; **Stena Line** *stenaline.com*; **Superfast Ferries** *superfast.com*; **Tallink & Silja Line (T&S)** *tallinksilja.com*; **Tirrenia** *tirrenia.it*; **Trasmediterránea** *trasmediterranea.es*; **TT-Line** *ttline.com*; **Ukrferry** *ukrferry.com*; **Unity Line** *unityline.eu*; **Venezia Lines** *venezialines.com*; **Ventouris Ferries** *ventourisferries.com*; **Viking Line** *vikingline.com*; **Virtu Ferries** *virtuferries.com*

ALBANIA

Trains HSH Albanian railways (no international trains) *hsh.com.al*
Ferries Greece Saranda–Corfu (Finikas); **Italy** Durrës to Bari, Ancona & Trieste (Adria, GNV, Ventouris)

AUSTRIA

Ecotourism Alpenverein Österreich Austrian Alpine Association *alpenverein.at*
Trains ÖBB Austrian railways; Railjet high-speed trains, Nightjet motorail sleepers and scenic lines (e.g. Arlberg, Semmering, Brenner) *oebb.at, nightjet.com*; **Westbahn** Vienna–Salzburg train *westbahn.at*
International trains Direct to France, Czech Republic, Germany, Hungary, Italy, Romania, Turkey, Ukraine, Russia & Switzerland. ÖBB trains from Vienna include Railjet to Budapest & Prague; Nightjet to Berlin, Brussels, Milan, Rome & Zürich; Motorail to Düsseldorf & Hamburg. Other sleepers from Vienna: CFR Dacia to Bucharest; PKP Chopin to Warsaw; RZD to Moscow and Nice.

BELGIUM

Ecotourism GR Walkers' network and GR long-distance trails *groteroutepaden.be, grsentiers.org*
Trains SNCB/NMBS Belgian railways *belgianrail.be*
International trains Direct to Austria, France, Germany, Luxembourg, the Netherlands & UK, e.g. from Brussels to Paris (Thalys), Cologne (DB ICE), London (Eurostar) and Vienna (ÖBB Nightjet sleeper)
Ferries England Zeebrugge–Hull (P&O)

BOSNIA & HERZEGOVINA

Trains ZFBH & ZRS National railways (no international trains) *zfbh.ba, zrs-rs.com*

BULGARIA

Trains BDZ Bulgarian railways *bdz.bg*
International trains Direct to Greece, Hungary, Romania & Turkey e.g. Sofia–Istanbul sleeper (TCDD)
Ferries Black Sea Burgas/Varna to Ukraine & Georgia (Bulgaria FF, Navibulgar, PBM)

CROATIA

Trains HŽ Croatian railways; Zagreb–Split sleeper *hzpp.hr*

International trains Direct to Austria, Germany, Hungary, Slovenia & Switzerland e.g. sleepers from Zagreb to Salzburg, Munich, Innsbruck & Zürich (HŽ); Split–Budapest (MÁV); Rijeka–Prague (RegioJet)
Ferries Croatia Ferries Guide to domestic routes, e.g. Dubrovnik, Split, Zadar & Rijeka to the Dalmatian islands *croatiaferries.com*; **Italy** Venice, Trieste, Ancona & Bari (GNV, Jadrolinija, Liberty, Venezia)

CYPRUS
Ferries Turkey Kyrenia–Taşucu (Akgünler Denizcilik)

CZECH REPUBLIC
Buses & trains České Dráhy Czech railways *cd.cz*; **RegioJet** Trains and buses *regiojet.com*
International trains Direct to Austria, Germany, Hungary, Poland, Russia & Slovakia, e.g. Prague to Vienna & Munich (ČD); sleepers to Budapest, Warsaw, Moscow, Bratislava, Košice and Rijeka (ČD, RegioJet, RZD)

DENMARK
Ecotourism Aktiv Danmark Cyclist-friendly accommodation (Bed + Bike), hiking routes (Walking Denmark), foodie tourism (Gastronomy Denmark) *www.activdanmark.com*; **Visit Denmark** Info on cycling, ferries, food & design *visitdenmark.com*
Trains DSB Danish railways *dsb. dk*; **International trains** Direct to Germany & Sweden, e.g. Copenhagen to Hamburg, Berlin, Malmö & Stockholm (DB ICE, SJ, Snälltåget sleeper)
Ferries Domestic (Scandlines); **Germany** Rødby–Puttgarden, Gedser–Rostock (Scandlines); **Faroe Is & Iceland** from Hirtshals (Smyril); **Norway** Copenhagen & Frederikshavn to Oslo (DFDS); Hirtshals to Larvik, Langesund, Kristiansand, Stavanger & Bergen (Fjord Line, Color); **Poland** Copenhagen–Świnoujście (Polferries); **Sweden** Frederikshavn–Göteburg; Grenaa–Halmstad (Stena)

ESTONIA
Trains EVR Estonian railways *evr.ee*; **International trains** Direct to Russia (RZD Tallinn–Moscow)
Ferries From Tallinn: **Finland** Helsinki & Hanko (Eckerö, Moby SPL, Viking, T&S); **Russia** St Petersburg (Moby SPL); **Sweden** Stockholm (DFDS, Moby SPL)

FINLAND
Ecotourism Nordic Swan Eco-certification scheme *joutsenmerkki.fi*
Trains VR Finnish railways, e.g. Helsinki–Lapland Santa Claus Express sleeper *vr.fi*; **International trains** Direct to Russia (RZD Helsinki-Moscow)
Ferries Domestic Helsinki & Turku to Åland (Viking); **Estonia** Helsinki & Hanko to Tallinn (DFDS, Eckerö, Moby SPL, T&S, Viking); **Germany** Helsinki-Travemünde (Finnlines); **Latvia** Helsinki–Riga (T&S); **Russia** Helsinki-St Petersburg (Moby SPL); **Sweden** Helsinki, Turku & Åland to Stockholm; Åland & Naantali to Kapellskär (Finnlines, Moby SPL, T&S, Viking)

FRANCE
Ecotourism Balades Paris Durable Self-guided walking tours in Paris *baladesparisdurable.fr*; **Bike About Tours** Cycle tours in Paris, Versailles & Champagne *bikeabouttours.com*; **FF Randonnée** GR long-distance walking trails *www.ffrandonnee.fr*; **France Écotours** Gypsy-style caravan trips, treehouses, cycling, hiking & vegan tours *france-ecotours.com*; **Nicols** Canal boat holidays in France, Germany & Portugal *www.boat-renting-nicols.co.uk*; **Pelagos Sanctuary** Ligurian Sea (France/Monaco/Italy) *sanctuaire-pelagos.org*
Buses & trains Altibus Buses to French Alps resorts *altibus.com*; **SNCF** French railways; TGV InOui & Ouigo high-speed trains *sncf.com*; **Trinighellu** Corsican train *train-corse. com*
International trains Direct to Belgium, Germany, Italy, the Netherlands, Portugal, Russia, Spain, Switzerland & UK, e.g. Paris–Frankfurt (DB ICE); Paris–Venice (Thello sleeper); Lyon–Turin (SNCF TGV); Marseille–Milan (Thello); Paris–Amsterdam (Thalys); Paris & Nice to Moscow (RZD sleeper); Paris–Barcelona (Renfe/SNCF TGV); Dijon to Basel & Zürich (TGV Lyria); Paris, Avignon and the French Alps to London (Eurostar); Calais-Folkestone motorail (Eurotunnel)
Ferries Domestic Marseille, Toulon & Nice to Corsica (Corsica Ferries, Corsica Linea, La Méridionale); **Balearic Islands** Toulon to Mallorca & Menorca (Corsica Ferries); **Channel Islands** from N France (Condor, Manche Îles); **Ireland** Roscoff & Cherbourg to Dublin, Cork & Rosslare (Brittany, Irish, Stena); **Italy** Corsica to Liguria & Tuscany; Corsica, Marseille, Toulon & Nice to Sardinia (Blu Navy, Corsica Ferries, Corsica Linea, La Méridionale, Moby Lines); **North Africa** Marseille & Sète to Morocco, Algeria & Tunisia (Corsica Linea, CTN, Grimaldi); **UK** Calais-Dover (P&O); N France to S England (Brittany)

GERMANY
Ecotourism ADFC Cycling organization *adfc.de*; **Bett und Bike** Cyclist-friendly accommodation *bettundbike.de*; **Deutscher Wanderverband** Walkers' association *wanderverband.de*
Trains Deutsche Bahn (DB) German railways; ICE high-speed trains *bahn.com*; **FlixTrain** Low-cost trains *flixtrain.com*; **Harz Narrow-Gauge Railway** *hsb-wr.de*; **Urlaubs-Express** Motorail sleepers *urlaubs-express.de*
International trains Direct to Austria, Belgium, Croatia, Czech Republic, Denmark, France, Hungary, Italy, the Netherlands, Poland, Russia, Sweden & Switzerland, e.g. Berlin-Vienna (ÖBB Nightjet); Cologne-Brussels (Thalys); Munich-Zagreb (HŽ sleeper); Munich-Prague (ČD); Berlin & Hamburg to Copenhagen, Malmö & Stockholm (Snälltåget sleeper); Frankfurt-Paris (DB ICE); Munich-Budapest (MÁV sleeper); Hamburg-Verona (Urlaubs motorail); Berlin-Warsaw (PKP); Berlin-Moscow (RZD); Kiel-Basel (DB sleeper)
Ferries Denmark Puttgarden–Rødby, Rostock-Gedser (Scandlines); **Finland** Travemünde-Helsinki (Finnlines); **Latvia** Travemünde-

Liepãja (Stena); **Lithuania** Kiel–Klaipėda (DFDS); **Norway** Kiel–Oslo (Color); **Sweden** Kiel–Göteborg; Travemünde & Rostock to Mälmo/Trelleborg (DFDS, Finnlines, Stena, TT)

GREECE
Ecotourism Tethys Whale and dolphin conservation trips, Ionian Sea *whalesanddolphins.tethys.org*
Trains TrainOSE Greek railways e.g. Athens to Thessaloniki & Patras *trainose.gr*; **International trains** Direct to Bulgaria, North Macedonia & Serbia e.g. Thessaloniki–Sofia (TrainOSE), Thessaloniki–Belgrade (SV)
Ferries From the mainland ports of Pireas, Rafina, Igoumenitsa, Patras & Thessaloniki and the Greek Islands, to:
Greek Islands Anek Lines *anek.gr*; Blue Star Ferries *bluestarferries.com*; Erturk Lines *erturk.com.tr*; Golden Star Ferries *goldenstarferries.gr*; Hellenic Seaways *hellenicseaways.gr*; Levante Ferries *levanteferries.com*; Minoan Lines *minoan.gr*; Seajets *seajets.gr*; Zante Ferries *zanteferries.gr*
International Albania Corfu–Saranda (Finikas); **Italy** Bari, Ancona, Venice etc. (Anek, Grimaldi, Liberty, Minoan, Superfast, Ventouris); **Turkey** Aylavik, Bodrum, Fethiye, Marmaris etc. (Dodecanese FD, Erturk, Sea Dreams)

HUNGARY
Ecotourism Green Guide Budapest *greenguide.hu*
Trains GySEV Hungarian-Austrian railways *gysev.hu*; **MÁV** Hungarian railways *mavcsoport.hu*; **International trains** Direct to Austria, Croatia, Czech Republic, Germany, Poland, Romania, Slovakia, Slovenia, Switzerland & Ukraine, e.g. sleepers from Budapest to Bucharest, Munich, Zürich and the Adriatic Express to Zagreb & Split (MÁV)

ICELAND
Ecotourism Nordic Swan Eco-certification scheme *svanurinn.is*
Ferries Denmark & Faroe Islands Seyðisfjörður to Tórshavn & Hirtshals (Smyril Line *smyrilline.com*)

IRELAND
Ecotourism Clissmann Horse Caravans Horse and donkey holidays, County Wicklow *clissmannhorsecaravans.com*;
Walking Ireland Guided and self-guided tours *walkingireland.ie*
Buses & trains Bus Eireann Intercity buses *buseireann.ie*; **Iarnród Éireann / Irish Rail** *irishrail.ie*; **International trains** Direct to Northern Ireland, e.g. Dublin–Belfast (IÉ/NI)
Ferries Spain Rosslare–Bilbao (Brittany); **France** Dublin, Cork & Rosslare to Roscoff & Cherbourg (Brittany, Irish, Stena); **UK, England** Dublin–Liverpool (P&O); **UK, Wales** Dublin–Holyhead, Rosslare–Fishguard, Rosslare–Pembroke (Irish, Stena); **UK, Isle of Man** from Dublin (Steam Packet)

ITALY
Ecotourism Agriturismo Rural accommodation and farmstays *agriturismo.it*; **FIE** Walkers' network *fieitalia.com*; **Tethys** Whale and dolphin conservation trips, Ligurian Sea *whalesanddolphins.tethys.org*
Trains Italo High-speed trains *italotreno.it*; **Trenitalia** Italian railways; Frecce high-speed trains and Intercity Notte sleepers (e.g. Milan–Palermo, via Messina train-ferry) *trenitalia.com*; **Trenord** Trains, Lombardy *trenord.it*; **International trains** Direct to Austria, France, Germany, Russia & Switzerland, e.g. Rome, Florence, Venice, Milan & Verona to Vienna (ÖBB Nightjet); Milan–Marseille (Thello), Venice–Paris (Thello sleeper), Verona–Munich (DB); Milan–Moscow (RZD sleeper); Milan–Geneva (SBB/Trenitalia)
Ferry operators Adria Ferries (Adr) *adriaferries.com*; **Anek Lines** (Anek) *anek.gr*; **Blu Navy** (Blu) *blunavytraghetti.com*; **Caronte & Tourist** (C&T) *carontetourist.it*; **Corsica Ferries** (CorF) *corsica-ferries.com*; **Corsica Linea** (CorL) *corsicalinea.com*; **Grandi Navi Veloci** (GNV) *gnv.it*; **Grimaldi Lines** (Gri) *grimaldi-lines.com*; **Gruppo Armatori Garganici** (GAG) *navitremiti.com*; **Jadrolinija** (Jad) *jadrolinija.hr*; **Liberty Lines** (Lib) *libertylines.it*; **La Méridionale** (Mér) *lameridionale.com*; **Minoan Lines** (Min) *minoan.*

gr; **Moby Lines** (MobL) *mobylines.com*; **Navigazione Libera del Golfo** (NLG) *navlib.it*; **Superfast Ferries** (Sup) *superfast.com*; **Toremar** (Tore) *toremar.it*; **Tirrenia** (Tirr) *tirrenia.it*; **Venezia Lines** (Vene) *venezialines.com*; **Ventouris Ferries** (Vent) *ventourisferries.com*; **Virtu Ferries** (Vir) *virtuferries.com*
Domestic ferries Arcipelago Campano from Campania (GAG, NLG); **Sardinia, Sicily & nearby islands** from Liguria, Tuscany, Rome, Naples & Calabria (Blu, C&T, CorF, GNV, Gri, Lib, MobL, Tirr); **Sicily** from Sardinia (Tirr); **Arcipelago Toscano** from Tuscany (Blu, CorF, Tor, MobL); **Isole Tremiti** from Molise & Puglia (GAG, NLG, Tirr)
International ferries run throughout the western and central Mediterranean region (except to mainland France and the Balearic Islands); **Albania** from Bari, Ancona & Trieste (Adr, GNV, Vent); **Corsica** from Elba & Livorno (CorFer, MobL); **Corsica** from Sardinia (Blu, CorF, CorL, Mér); **Croatia & Slovenia (Istria)** from Venice, Trieste, Ancona & Bari (GNV, Jad, Lib, Vene); **France** Sardinia–Marseille (CorL); **Greece** from Puglia, Ancona & Venice (Anek, Gri, Lib, Min, Sup, Vent); **Malta** from Sicily & Salerno (Gri, Vir); **Montenegro** Bari–Bar (Jad); **North Africa** Liguria, Rome, Salerno & Sicily to Nador, Tangier & Tunis (CTN, Gri, GNV, Tirr); **Spain** Liguria, Rome & Sardinia to Barcelona (GNV, Gri)

LATVIA
Trains & ferries LDz Latvian railways *ldz.lv*; **International trains** Direct to Russia (RZD Riga–Moscow); **Ferries to Finland, Germany & Sweden** Riga–Helsinki, Liepāja–Travemünde, Riga–Stockholm, Ventspils–Nynäshamn (Stena, T&S)

LITHUANIA
Trains & ferries LG Lithuanian railways *litrail.lt*; **International trains** Direct to Latvia & Poland; **Ferries to Germany & Sweden** Klaipėda to Kiel, Karlshamn & Trelleborg (DFDS, TT)

LUXEMBOURG
Trains CFL Luxembourg railways *cfl.lu*; **International trains** Direct to Belgium, France & Germany (SNCF, DB)

MALTA
Ferries Domestic Valletta Ferry Services *vallettaferryservices.com*; **Italy** Valletta–Salerno, Valletta–Sicily (Grimaldi, Virtu)

MONTENEGRO
Trains & ferries ŽCG Montenegro railways *zcg-prevoz.me*; **International trains** Direct to Serbia, e.g. Bar–Belgrade (SV); **Ferries to Italy** Bar–Bari (Jadrolinija)

THE NETHERLANDS
Ecotourism Fietsers Welkom! Cyclist-friendly accommodation *allefietserswelkom.nl*; **Holland Cycling Routes** Tourism information *hollandcyclingroutes. com*; **Natuurkampeerterreinen** Natural campsites in rural areas *natuurkampeerterreinen.nl*; **Stayokay** Hostelling International *stayokay. com*; **Trekkershutten** Hikers' huts *trekkershutten.nl*; **Vrienden op de Fiets** Homestays for cyclists *vriendenopdefiets.nl*; **Wandelnet** Walkers' network and GR long-distance trails *wandelnet.nl*
Trains & ferries NS Netherlands railways *ns.nl*; **International trains** Direct to Belgium, France, Germany & UK, e.g. Rotterdam-Brussels & Amsterdam–Paris (Thalys), Amsterdam–Düsseldorf (DB ICE), Amsterdam–London (Eurostar); **Ferries to England** Amsterdam–Newcastle (DFDS); Hoek van Holland–Harwich (Stena); Rotterdam–Hull (P&O)

NORTH MACEDONIA
Trains MŽ North Macedonia railways *mzi.mk*; **International trains** Direct to Greece & Serbia

NORWAY
Ecotourism Brim Explorer Arctic wildlife-watching by eco-boat *brimexplorer.com*; **Nordic Swan** Eco-certification scheme *svanemerket.no*
Trains The Flåm Railway and **Norway In A Nutshell** Tourist route *visitflam.*

com; **Vy** Norwegian buses and railways; sleeper trains from Oslo to Stavanger, Bergen, Trondheim & Bodø *vy.no*; **International trains** Direct to Sweden, e.g. Oslo to Stockholm & Göteborg (SJ)
Domestic ferries Fjord1 *fjord1.no*; Fjord Line *fjordline.com* (e.g. Bergen to Stavanger); Norled *norled.no* (e.g. Bergen to Flåm); Rødne Fjord Cruise *rodne.no*
International ferries Denmark Oslo to Copenhagen & Frederikshavn (DFDS); Larvik, Langesund, Kristiansand, Stavanger & Bergen to Hirtshals (Color, Fjord Line); **Germany** Oslo–Kiel (Color); **Sweden** Sandefjord–Strømstad (Color)

POLAND
Buses, trains & ferries Agat International buses *agat.eu*; **PKP** Polish railways; Gdansk-Warsaw-Krakow high-speed trains *pkp.pl*, *intercity.pl*; **Polregio** Polish regional railways *polregio.pl*; **International trains** Direct to Austria, Czech Republic, Germany & Russia, e.g. express from Warsaw to Berlin and sleepers to Vienna, Prague and Moscow (PKP, RZD); **Ferries to Denmark** Świnoujście-Copenhagen (Polferries); **Ferries to Sweden** Gdańsk, Gdynia & Świnoujście to Karlskrona, Nynäshamn, Trelleborg & Ystad (Polferries, Stena, TT, Unity)

PORTUGAL
Ecotourism Douro Azul River cruises *douroazul.com*; **FCMP** Outdoor activities and walking trails *fcmportugal.com*
Trains Comboios de Portugal (CP) Portuguese railways; Porto-Lisbon-Faro high-speed train *cp.pt*; **International trains** Direct to Spain, e.g. Lisbon–Madrid Lusitania Trenhotel (Renfe), Porto–Vigo Celta (CP)

ROMANIA
Trains & ferries CFR Călători Romanian railways *cfrcalatori.ro*; **International trains** Direct to Austria, Bulgaria, Hungary & Turkey, e.g. Bucharest to Sofia (CFR), Budapest & Vienna (CFR sleeper) and Istanbul (TCDD sleeper); **Ferries to Turkey** Constanța-Karasu (Sea Lines)

RUSSIA
Trains RZD Russian Railways; Moscow–St Petersburg Sapsan high-speed train *rzd.ru*
International trains Direct to Europe & Asia. St Petersburg–Helsinki high-speed train; long-distance sleepers from Moscow to Helsinki, Tallinn, Riga, Kyiv, Odesa, Warsaw, Prague, Milan, Vienna, Nice, Paris, Berlin and Frankfurt; Trans-Siberian/Mongolian/Manchurian railway from Moscow to Vladivostok, Ulaanbaatar & Beijing (RZD).
Ferries to South Korea & Japan Vladivostok to Donghae & Sakaiminato (DBS Cruise Ferry *dbsferry.com*); **Ferries to Baltic ports** St Petersburg to Helsinki, Stockholm & Tallinn (Moby SPL)

SERBIA
Trains Srbija Voz Serbian railways *srbvoz.rs*; **International trains** Direct to Austria, Bulgaria, Greece, Montenegro, Switzerland & Turkey, e.g. Belgrade–Bar Montenegro Express & Belgrade–Thessaloniki Hellas Express (SV)

SLOVAKIA
Trains ZSSK Slovak railways *zssk. sk*; **International trains** Direct to Czech Republic, Germany, Hungary, Poland, Slovakia & Ukraine, e.g. Košice–Prague (RegioJet sleeper), Bratislava–Berlin (MÁV Metropol sleeper), Bratislava–Budapest (ČD) and Košice–Mukachevo (ZSSK)

SLOVENIA
Ecotourism Dolina Soče Soča Valley and river *soca-valley.com*
Trains & ferries SŽ Slovenian railways *slo-zeleznice.si*; **International trains** Direct to Austria, Croatia, Germany, Hungary, Serbia & Switzerland, e.g. Ljubljana to Zagreb (SŽ) and to Salzburg, Munich, Innsbruck & Zürich (HŽ sleeper); **Ferries to Italy** Piran to Trieste & Venice (Liberty, Venezia)

SPAIN

Ecotourism Casas Rurales Rural villas, guesthouses and farmstays *casasrurales.net*; **Association of Ecotourism in Spain** Holidays and activities *ecotouristinspain. com*; **FEDME** GR long-distance walking trails *fedme.es*; **Santiago de Compostela Pilgrim's Reception Office** *oficinadelperegrino.com*; **Vías Verdes** Walking and cycling paths on disused railways *viasverdes.com*

Buses & trains Alsa Long-distance buses *alsa.com*; **Renfe** Spanish railways; ACE and Alvia high-speed trains, Trenhotel sleepers and luxury hotel-trains (Al Ándalus, Costa Verde Express, El Expreso de La Robla, Transcantábrico Gran Lujo) *renfe. com*; **International trains** Direct to France & Portugal, e.g. Barcelona-Paris (SNCF TGV); Madrid-Lisbon Lusitania and Irún/San Sebastian-Lisbon Surexpresso Trenhotel sleepers (Renfe); Vigo-Porto Celta (CP)

Domestic ferries Balearic Is from Barcelona, Dénia & Valencia (Baleària); **Balearic Is** inter-island (Aqua Bus, Baleària, Corsica Ferries, Med Pitiusa, Trasmed); **Canary Is** from Cadiz & Huelva (Baleària, F Olsen, FRS, Trasmed); **Canary Is** inter-island (F Olsen, Nav Armas, Trasmed)

International ferries France Mallorca/Menorca-Toulon (Corsica Ferries); **Ireland** Bilbao-Rosslare (Brittany); **Italy** Barcelona to Liguria, Rome & Sardinia (GNV, Grimaldi); **Morocco, Spanish North Africa & Algeria** from Barcelona & Andalucía, e.g. Tarifa-Tanger (AML, Baleària, FRS, GNV, Grimaldi, Inter, Nav Armas, Trasmed); **UK** Bilbao & Santander to Plymouth & Portsmouth (Brittany)

SWEDEN

Ecotourism Archipelago Foundation Stockholm Archipelago *archipelagofoundation.se*; **Nordic Swan** Eco-certification scheme *svanen.se*; **Skåneleden** Walking trail, Scania *skaneleden.se*

Trains Inlandsbanan Kristinehamn-Gällivare *res.inlandsbanan.se*; **MTRX** Stockholm-Göteborg *mtrx. travel*; **Snälltåget** Stockholm-Jämtland sleepers *snalltaget.se*; **SJ** Swedish railways; high-speed trains and sleepers from Stockholm, Göteborg & Malmö to N Sweden *sj.se*; **International trains** Direct to Denmark, Germany & Norway, e.g. Stockholm & Göteborg to Oslo; Stockholm & Malmö to Copenhagen, Hamburg & Berlin; Östersund-Trondheim (SJ, Snälltåget)

Domestic ferries Gotland Nynäshamn & Oskarshamn to Visby (Destination Gotland *destinationgotland.se*); **Stockholm Harbour & Archipelago** (Viking Line *vikingline.com*; Stromma *stromma.com*; Waxholmsbolaget *waxholmsbolaget.se*)

International ferries Denmark Göteburg-Frederikshavn; Halmstad-Grenaa; Helsingborg-Helsingør (Stena, Scandlines); **Estonia** Stockholm-Tallinn; Kapellskär-Paldiski (DFDS, Moby SPL, T&S); **Finland** Stockholm & Kapellskär to Åland, Naantali, Turku & Helsinki (Finn, Moby SPL, T&S, Viking); **Germany** Göteborg-Kiel; Mälmo & Trelleborg to Travemünde & Rostock (DFDS, Finnlines, Stena, TT); **Latvia** Stockholm-Riga; Nynäshamn-Ventspils (Stena, T&S); **Lithuania** Karlshamn-Klaipėda (DFDS, TT); **Norway** Strømstad-Sandefjord (Color, Fjord Line); **Poland** Karlskrona-Gdynia; Trelleborg-Swinoujście; Ystad-Świnoujście; Nynäshamn-Gdańsk (Polferries, Stena, TT, Unity); **Russia** Stockholm-St Petersburg (Moby SPL)

SWITZERLAND

Ecotourism Schweizer Wanderwege Swiss hiking trails *wandern.ch*

Trains Bernina Express *berninaexpress.ch*; **Glacier Express** *glacierexpress.ch*; **SBB CFF FFS** Swiss railways *sbb.ch*; **Rhaetian Railway** *rhb.ch*; **International trains** Direct to Austria, Croatia, France, Germany, Hungary, Italy & Slovenia, eg Lausanne-Paris (TGV Lyria), Zürich-Freiburg (ÖBB), Zürich-Hamburg (ÖBB Nightjet), Geneva-Milan (SBB/Trenitalia)

TURKEY

Trains TCDD Turkish railways, e.g. Istanbul-Ankara high-speed train *tcdd.gov.tr*; **International trains** Direct to Austria, Bulgaria, Iran, Romania & Serbia, e.g. Istanbul-Bucharest Bosphor Express; Istanbul-Sofia Sofia Express; Edirne-Villach Optima Express; Ankara-Tehran Transasia Express (TCDD)

Ferries Cyprus Taşucu-Kyrenia (Akgünler Denizcilik); **Greek Islands** from Aylavik, Bodrum, Fethiye, Marmaris etc. (Dodecanese FD, Erturk, Sea Dreams); **Romania** Karasu-Constanţa (Sea Lines); **Ukraine** Karasu-Odesa (Ukrferry, Sea Lines)

UKRAINE

Ecotourism Association for the Promotion of Rural Green Tourism Homestay network *greentour.com.ua*

Trains & ferries Ukrainian Railways *uz.gov.ua*; **International trains** Direct to Austria, Hungary, Russia & Slovakia, e.g. Kyiv-Budapest-Vienna (UZ sleeper); Odesa-Kyiv-Moscow (RZD sleeper), Mukachevo-Košice (ZSSK); **Black Sea ferries** (Odesa to Bulgaria, Turkey & Georgia) Navibulgar, Ukrferry, Sea Lines

UNITED KINGDOM

Ecotourism The Landmark Trust Holidays in historic buildings *landmarktrust.org.uk*; **National Trails** Long-distance walking, cycling and bridle routes, England & Wales *nationaltrail.co.uk*; **Scotland's Great Trails** Long-distance walking and riding routes, e.g. West Highland Way *scotlandsgreattrails.com*; **Sustrans** National Cycle Network *sustrans.org.uk*; **Under The Thatch** Restored cottages, Wales & Ireland *underthethatch.co.uk*; **Wales Coast Path** *walescoastpath.gov. uk*; **Wildlife Trust of South & West Wales** Mainland and island reserves *welshwildlife.org*

Buses & trains Caledonian Sleeper London–Scotland night trains *sleeper.scot*; **GWR** Trains from London to Wales & West Country; London–Penzance Night Riviera sleeper *gwr.com*; **National Express** Long-distance buses *nationalexpress. com*; **National Rail Enquiries** *nationalrail.co.uk*; **ScotRail** Scottish trains; West Highland Line *scotrail. co.uk*; **Translink** Transport in N Ireland *translink.co.uk*; **Traveline** Public transport information *traveline.info*; **Traveline Cymru** Transport in Wales *traveline.cymru*; **Traveline Scotland** Transport in Scotland *travelinescotland.com*; **International trains** direct to France, Belgium & the Netherlands (Eurostar, e.g. London to Paris, Marseille, French Alps, Brussels, Rotterdam & Amsterdam); motorail to France (Eurotunnel, Folkestone–Calais) **Domestic ferries Channel Islands** from Poole & Portsmouth (Condor Ferries *condorferries.co.uk*); **Inner & Outer Hebrides** from west Scotland (Calmac Ferries *calmac.co.uk*); **Isle of Man** from Merseyside & Heysham (Steam Packet *steam-packet.com*); **Northern Ireland** from Liverpool & Cairnryan (P&O Ferries *poferries. com*, Stena Line *stenaline.com*); **Orkney & Shetland** from Aberdeen & Caithness (Northlink Ferries *www. northlinkferries.co.uk*, Pentland Ferries *pentlandferries.co.uk*); **Scilly** from Penzance (Scillonian Ferry *islesofscilly-travel.co.uk*); **Isle of Wight** from Hampshire (Red Funnel *redfunnel.co.uk*, Wightlink *wightlink. co.uk*) **International ferries Belgium & the Netherlands** Hull–Zeebrugge; Hull–Rotterdam; Newcastle–Amsterdam; Harwich–Hoek van Holland (DFDS, P&O, Stena); **France** Newhaven–Dieppe (DFDS), Dover–Calais (P&O), Plymouth, Poole, Portsmouth & Channel Islands to NW France (Brittany, Condor, Manche Îles); **Ireland** Liverpool, Holyhead & Isle of Man to Dublin (Irish, P&O, Steam Packet, Stena), Pembroke & Fishguard to Rosslare (Irish, Stena); **Spain** Plymouth & Portsmouth to Santander & Bilbao (Brittany)

AFRICA

Ecotourism Acacia Africa Overland tours *acacia-africa.com*; **African Trails** Overland tours *africantrails. co.uk*; **Africa's Finest** Sustainable safari lodges *africasfinest.co.za*; **Another Africa** Bespoke wildlife and cultural tours *anotherafrica.com*; **Expert Africa** Safaris in southern and East Africa *expertafrica.com*; **Fair Trade Tourism** Certified lodges & experiences *fairtrade.travel* **Buses & trains Baz Bus** Hop-on-hop-off backpacker route, South Africa *bazbus.com*; **The Blue Train** Luxury sleeper, Cape Town–Pretoria *www. bluetrain.co.za*; **ONCF** Moroccan railways *oncf.ma*; **Rovos Rail** Luxury trains, southern & East Africa, e.g. Shongololo Express *rovos.com*; **SNCFT** Tunisian railways *sncft.com.tn* **Ferries France** Tunisia & Algeria to Marseille; Morocco to Sète (CTN, Corsica Linea, GNV) **Italy** Morocco & Tunisia to Liguria, Rome, Salerno & Sicily (CTN, Corsica Linea, GNV, Grimaldi, Tirrenia); **Spain** Morocco, Spanish North Africa & Algeria to Andalucía & Barcelona (Africa Morocco Link, Baleària, FRS, GNV, Grimaldi, Inter, Naviera Armas, Trasmediterránea)

ASIA, AUSTRALASIA & PACIFIC

Ecotourism Asian Ecotourism Network *asianecotourism.org* **Ecotourism Australia** Certification scheme *ecotourism.org.au* **Buses & trains 12Go** Bus, train & ferry bookings, Southeast Asia *12go.asia*; **China DIY Travel** Railway bookings *china-diy-travel.com*; **China Railway** *12306.cn*; **The Ghan, Indian Pacific, Overland & Great Southern** Long-distance railways, Australia *journeybeyondrail.com. au*; **Greyhound Australia** Long-distance buses *greyhound.com.au*; **Japan Bus Online** Booking service *japanbusonline.com*; **Japan Rail Pass** Covers trains, JR buses & Miyajima Ferry *japanrailpass.net*; **NSW TrainLink** Trains & buses, SE Australia *transportnsw.info*; **Queensland Railways** Trains, NE Australia *queenslandrail.com.au*; **RZD** Trans-Manchurian and Trans-Mongolian railway to Moscow from China & Mongolia *rzd.ru* **Ferries ASCO** Caspian Sea, Baku (Azerbaijan) to Turkmenbashi (Turkmenistan) & Aktau (Kazakhstan) *asco.az*; **DBS Cruise Ferry** Between South Korea, Russia & Japan *dbsferry.com*; **Limbongan Maju** Malaysia to Singapore & Indonesia *limbonganmaju.com*; **Shanghai Ferry Co** China–Japan *shanghai-ferry.co.jp*; **Spirit of Tasmania** Melbourne–Tasmania, Australia *spiritoftasmania.com.au*

AMERICAS & ANTARCTICA Ecotourism IAATO Antarctica tourism association *iaato.org*; **National Park Service** USA *nps.gov*; **National Trails** Long-distance hiking and cycling, USA *americantrails.org*; **Parks Canada** *pc.gc.ca*; **Trans Canada Trail** *thegreattrail.ca* **Buses & trains Amtrak** Railways and rail passes, USA *amtrak.com*; **BoltBus** Intercity buses, N America *boltbus. com*; **FlixBus** Low-cost long-distance buses, USA *global.flixbus.com*; **Greyhound** Intercity buses, USA *greyhound.com*; **VIA Rail** Railways and rail passes, Canada *viarail.ca*. Long-distance buses connecting South American cities are best booked locally. **Ferries Alaska Marine Highway** USA *dot.alaska.gov/amhs*; **BC Ferries** Hybrid ferries, Canada *bcferries.com*; **L'Express des Îles** Lesser Antilles *express-des-iles.fr*; **Trinidad and Tobago Inter-Island Ferry Service** *ttitferry.com*; **Washington State Ferries** Hybrid ferries, USA *wsdot. wa.gov/ferries*

25 FLIGHT-FREE JOURNEYS IN EUROPE

BY DAY Fifteen trips of under 1,000 kilometres by train, ferry, bus or electric car

FROM	TO	FASTEST TRAIN OR FERRY	FASTEST JOURNEY TIME	DRIVING DISTANCE (KM)
Brussels	Cologne	Thalys	1hr 47min	224
London	Brussels	Eurostar	1hr 53min	367
Vienna	Budapest	ÖBB Railjet	2hr 20min	243
Madrid	Budapest	Renfe AVE	2hr 30min	642
Paris	Marseille	SNCF TGV	3hr 5min	775
Rome	Milan	Frecciarossa 1000	3hr 10min	537
Helsinki	St Petersburg	RZD Allegro	3hr 30min	391
Munich	Berlin	DB ICE	3hr 55min	585
London	Edinburgh	LNER	4hr 19min	637
Copenhagen	Hamburg	DB IC	4hr 36min	334
Copenhagen	Stockholm	SJ Snabbtåg	4hr 52min	657
Paris	Turin	SNCF TGV	5hr 28min	775
London	Lyon	Eurostar	5hr 41min	920
Liverpool (UK)	Dublin via Holyhead	TFWRail & Irish Ferries	5hr 45min	162
Paris	Girona	SNCF TGV	5hr 48min	942

BY NIGHT Ten classic trips by sleeper train, overnight ferry, long-distance bus or electric car

FROM	TO	OVERNIGHT TRAIN OR FERRY	TRAIN/FERRY JOURNEY TIME	DRIVING DISTANCE (KM)
Copenhagen	Oslo	DFDS ferry	19hr	603
Vienna	Berlin	ÖBB Nightjet	12hr 38min	685
Amsterdam	Bludenz	Alpen Express	16hr 27min	873
London	Inverness	Caledonian Sleeper train	11hr 27min	902
Paris	Venice	Thello train	14hr 10min	1,111
Portsmouth (UK)	Bilbao (Spain)	Brittany Ferries	23hr 45min	1,140
Milan	Palermo	Trenitalia Intercity Notte train	19hr 3min	1,470
Nice	Moscow	RZD Russian Railways train	47hr 15min	3,073
Huelva (Spain)	Las Palmas (Canary Islands)	Baleària, Fred Olsen or FRS ferry	34–36hr	n/a
Hirtshals (Denmark)	Seyðisfjörður (Iceland)	Smyril Line ferry	67hr	n/a

INDEX

A

Abraham Path 163
Ærø 111
Africa 208–13, 244
Al Andalus 65
Al Boraq 67, 69
Åland Islands 71, 111
Alaska 233, 239
Alba 173
Alpen Express 124
Alps 118–25, 156–9, 164–7
Alsace 154–5
Amsterdam 153, 155, 170, 209
Andalucía 65
Angers 41
Antarctica 234–9, 244
Arctic Circle 129, 233, 239
Argentina 236–7
Arlberg 122–5
Arran 77
Atlas Mountains 188–91
Australia 103, 107, 141, 215, 222–7,
    243, 244
Austria 58–61, 122–5, 159, 168–9, 197
Azores 215–16

B

Balearic Islands 90–1
Bali 221
Bar 133
Barcelona 235–6
Bardsey 81, 138
Bavaria 164–7, 170
Belgium 45, 150–3
Belgrade 133, 169–70
Bergen 111, 129, 151
Bergensbanen 129
Berlin 58–61, 171
Bernina Express 118–21
Bilbao 65
Black Forest 50–3, 169
Bludenz 124
Bodø 129
Bodrum 100–3
Borneo 244
Bosnia 187
Bratislava 61, 169
Brazil 236, 244
Bremerhaven 221
Brennerbahn 125
Brighton and Hove 37
Bristol 34–7
Brittany 86–9, 244
Bruges 151, 153
Brussels 45, 203
Bucharest 170
Budapest 61, 169–71
Buenos Aires 236
Bulgaria 130–3

C

Cambodia 200–1, 203
Canary Islands 99, 113, 181, 219–21
Cape Nordkinn 203
Cape Town 209–13, 215, 244
Caribbean 215, 219–21
Carpathian Mountains 61, 182–5
Casablanca 67–9

Ceredigion 89
Channel Islands 89
Charleston 244
Chernobyl 197
Chile 237
China 199, 200, 203, 207, 227,
    243, 244
Circum-Baikal Railway 206
Cleveland Way 145
Clonegall 149
Colombia 244
Copenhagen 46–9, 180, 243
Córdoba 65, 67
Corsica 94–5
Costa Brava 177
Costa Rica 12, 244
Courchevel 158
Crete 107
Croatia 101, 104–7
Czech Republic 171

D

Dalarö 110
Dalmatian Coast 104–7
Danube River 61, 153, 168–71
Delft 45
Den Bosch 42–5
Denmark 46–9, 110–11, 150–3,
    217, 243
Dinaric Alps 133
Djibouti 227
Douro River 171
Dresden 195
Dublin 54–5, 149

E

Edinburgh 37
Elafonisos 99
Elaphiti Islands 105
Elba 99
Elbe 167, 171
England 34–7, 82–5, 145, 149, 151
Ericeira 92–3
Essaouira 69, 112–13
Estonia 70–3, 217
EuroVelo 150–3

F

Falkland Islands 239
Faroe Islands 103
Fes 69
Finistère 86–9
Finland 71–2, 111, 129, 159, 181,
    217, 243
Fjord coast 111, 127, 230, 233
Flåmsbana 126–9
Formentera 90–1
France 38–41, 65, 86–9, 91, 94–5,
    149, 153, 154–9, 160–3, 187, 195,
    221, 244
Freiburg 50–3
Frisian Islands 152–3
Füssen 166–7
Fuxing Hao 227

G

Galway 56–7
Genoa 223, 236

Germany 50–3, 58–61, 116–17, 150–3,
    164–7, 169–70, 171, 195, 217, 221, 230
Ghent 45
Giglio 99
Glacier Express 121
Glasgow 57, 77, 144
Gloggnitz 125
Gower Peninsula 136–7, 140–1
Granada 65, 67
Great Glen Way 145
Greece 99, 100–3, 107
Greenland 228–33
Guernsey 89

H

The Hague 45, 153
Hajar Mountains 191
Hamburg 61, 151, 153, 230
Harz Narrow-Gauge Railway 116–17
Haut-Beaujolais 146–9
Hebrides 76–9
Helsinki 71–2, 111, 129, 243
Hendaye 65, 221
Hossegor 93
Hungary 61, 197
Hvar 106
Hydra 99

I

Iceland 23, 103, 181, 228–33
Île de Ré 91
Indonesia 221
Innsbruck 124
Ireland 54–7, 93, 145, 146–9
Irkutsk 206
Iron Curtain Trail 153
Ischia 99
Isle of Portland 85
Isle of Wight 85
Istanbul 200
Italy 95–9, 118–21, 125, 172–7, 222–7

J

Japan 163, 204–7, 227, 243
Jerez 65
Jordan Trail 191
Jungfraujoch 121

K

Karhunkierros Trail 181
Karlsruhe 51
Kazakhstan 200, 203
Kiev 194–7
King Ludwig Way 164–7
Koblenz–Trier Railway 117
Kornati archipelago 105
Korona Express 130–3
Koufonisia 99
Krakow 195
Kumano Kodo 163
Kyoto 206–7

L

La Réunion 227
Laos 201, 203
Lapland 129
Latvia 73
Laugavegur Trail 181

Lech Zürs 125
Les Menuires 158
Lev Tolstoy 73
Liechtenstein 125
Lisbon 62–5
Llyn Peninsula 141
Lofoten Islands 28, 103, 221, 233
London 37
Lucerne 121
Lviv 195

M
Madrid 62–4
Malaysia 203
Malmö 179–80
Malta 95, 98–9
Marrakech 67–9, 197
Mauritius 227
Mekong River 201
Méribel 158
Milos 99
Montenegro 101, 103, 133
Morocco 66–9, 112–13, 188–91,
    197, 213
Moscow 73, 203, 205, 227, 243
Münster 53
Mürzzuschlag 125

N
Nairobi 213
Namibia 213, 244
Nancy 155
Nantes 41
Nesebar 131
Netherlands 42–5, 150–3, 155
New Zealand 227, 244
Nightjet 59–60
Nordland Line 129
North Sea Cycle Route 150–3
Norway 28, 49, 103, 111, 126–9, 151,
    221, 228–33, 239, 243

O
Oman 191, 227
Orkney 81, 103
Osaka 206
Oslo 49, 129, 221, 243
Osnabrück 53
Otter Trail 141

P
Painters' Way 167
Palatinate Wine Trail 167
Panama Canal 227
Paris 38–41, 195
Pelagiem Islands 99
Perm 204, 206
Peru 244
Petra 191
Piatra Craiului 185
Picos de Europa 65
Piedmont 172–5
Poland 71, 195, 197, 244
Ponza 99
Portugal 62–5, 92–3, 160–3,
    171, 214–17
Prague 171
Procida 99

Punta de Tarifa 203
Pyrenees 159

R
Rabat 69
Rallarvegen 129
Reykjavík 230–1, 233
Rhodes 101
Rhodope Mountains 130–3
Riga 73
Romania 153, 182–5, 244
Rossnowlagh 93
Rotterdam 45
Route 66 203
Russia 71, 72–3, 203, 206, 217, 243

S
St Anton 124, 159
St Moritz 121
St Petersburg 71, 72–3
Sakaiminato 206–7, 227
Samarkand 200
San Francisco 242
San Sebastián 65
Santa Claus Express 129
Santander 65
Santiago de Compostela 65, 160–3
Santillana del Mar 65
Sardinia 95, 96–7
Saxon Steam Railway 117
Scilly Isles 82–5
Scotland 37, 57, 76–9, 81, 103, 107,
    142–5, 151
Semmeringbahn 125
Serbia 133, 170
Seville 65
Shetland 103, 107, 151
Shinkansen 206
Siberia 204–7
Sicily 95, 97
Singapore 196–203, 221, 227,
    243, 244
Skåneleden 178–81
Skokholm 81, 138
Skomer 81, 138
Slovakia 61
Slovenia 186–7
Soča River 186–7
Sofia 131–2
South Africa 141, 209–13, 244
South Georgia 239
South Korea 207
Spain 62–4, 65–7, 90–1, 113,
    160–3, 177
Sri Lanka 227
Stockholm 49, 71, 243
Stockholm archipelago 108–11
Sud Expresso 65
Suez Canal 222–7
Svalbard 239
Svartsö 110
Sweden 49, 71, 108–11, 178–81,
    217, 243
Switzerland 118–21, 125, 197
Sylt 91

T
Tallinn 70–3
Tangier 66–7, 213
Ţarcu Mountains 185
Tarifa 113, 213
Thailand 201, 203
Tierra del Fuego 236–7, 244
Tirano 119, 121
Tokyo 204–7, 243
Top of Europe Railway 121
Trans-Manchurian Railway 203,
    207, 227
Trans-Mongolian Railway 203,
    207, 227
Trans-Siberian Railway 203,
    204–7, 227
Transcantábrico Gran Lujo 65
Transylvania 182–5
Tremitis 99
Trier 117
Trinighellu 95
Tromsø 233
Trondheim 129
Turkey 100–3, 200
Turku 71, 111

U
Ukraine 194–7
Ulaanbaatar 207
Ulster Way 145
United States 203, 214–17, 227,
    233, 242, 244
Uzbekistan 200

V
Val Thorens 156, 158–9
Vale of Eden 149
Vancouver 242, 244
Verona 125
Via Francigena 163
Vienna 58–61, 169–70
Vietnam 200–1, 203
Vitosha Express 133
Vladivostok 205, 206–7, 227
Vosges Mountains 154

W
Wales 81, 89, 136–41, 155
Warsaw 71
Wernigerode 116
West Highland Way 142–5
Whitsunday Islands 103
Wicklow 146–9

Y
Yokohama 227
York 57

Z
Zadar archipelago 105
Zaragoza 67
Zermatt 121
Zürich 125

ACKNOWLEDGEMENTS

The publishers would like to thank the following for supplying images.

Taku Bannai for the beautiful illustrations

All photographs © Shutterstock.com
4 Alex Stemmer; 7 wassiliy-architect; 12 Galyna Andrushko; 17 o.hudzeliak; 19 Tom Asz; 20 Petr Jilek; 23 Mikadun; 27 Dreamer Company; 28 Dominik Belica; 34 Paul D Smith; 36 Kollawat Somsri; 37 Sion Hannuna; 38 ivan bastien; 40 Kiev.Victor; 41 maziarz; 42 Petr Pohudka; 44 barmalini; 45 Celli07; 46 Iryna Kalamurza; 48 lauravr; 49; 50 Sina Ettmer Photography; 52 Sergey Dzyuba; 54 Evgeni Fabisuk; 56 Luca Fabbian; 57 ; 58 SP-Photo; 60 Mistervlad; 61 mRGB; 62 Sean Pavone; 64 Catarina Belova; 65 Marcin Krzyzak; 66 Boris Stroujko; 68 Balate Dorin; 69 posztos; 70 yari2000; 72 dimbar76; 73 irra_irra; 76, 79 Jaime Pharr; 79 below Fotimageon; 80 phildaint; 82 Neil Duggan; 84 Emily Luxton; 85 Andrew Roland; 86 andre quinou; 88 art_of_sun; 89 Rolf E. Staerk; 91 jotapg; 93 homydesign; 94 Andrea Sirri; 96 robertonencini; 97 Roman Sigaev; 98 Jaroslav Moravcik; 99 Konstantin Aksenov; 100 Sanchik; 102 leoks; 103 Andrew Mayovskyy; 104 xbrchx; 106 islavicek; 107 Roman Babakin; 108 NAN728; 110; 111 Camilo Torres; 113 Szymon Barylski; 117 dugdax; 118 Luca Santilli; 120 Estremo; 122 Umomos; 124 ErichFend; 126 Andrey Shcherbukhin; 128 In Green; 129 Krishna.Wu; 130, 218 S-F; 132 bretelky; 133 Dafinka; 136 Paul Cowan; 138 Ian Fletcher; 139 Arjen de Ruiter; 140 jax10289; 142 andy morehouse; 143 LouieLea; 146 Nataliia; 148 Alan Kraft; 149 Anton Sitnik; 150 Doris Oberfrank-List; 152 Doin; 153 emka74; 154 Oleksandr Osipov; 156 guruXOX; 158 nikolpetr; 159 as_trofey; 160 Jorge Rodriguez Lago; 162 gregorioa; 163 roberaten; 164 Rudy Balasko; 165 FooTToo; 166 Boryana Manzurova; 168 canadastock; 170 PavleMarjanovic; 171 INTERPIXELS; 172 Giorgio1978; 173 Alessandro Cristiano; 174 StevanZZ; 177 grutfrut; 178 Lena Si; 180 Tommy Alven; 181 Ingus Kruklitis; 182 Arpad Laszlo; 184 Giannis Papanikos; 187 marcin jucha; 188 romanmalik; 190 Ondrej Bucek; 191 Ryzhkov Oleksandr; 194 Oleg Totskyi; 196 Serhii Hrynkevych; 197 Pe3k; 198 Nattee Chalermtiragool; 200 michel arnault; 201 Jiranuch Suwanarat; 202 R. de Bruijn_Photography; 203 Efired; 204 My September; 206 Sezai Sen; 207 thipjang; 208, 242 Travel Stock; 210 Ondrej Prosicky; 211 Werner Lehmann; 212 Artush; 214 criben; 216 Subodh Agnihotri; 217 OneOfTheseDays83; 220 Nikiforov Alexander; 221 elvirkins; 222 RossHelen; 223 Mr.Gheith; 224 Anton_Ivanov; 226 Natsicha Wetchasart; 228 Vadim Petrakov; 231 Patpong Sirikul; 232 Vadim Petrakov; 233 Beata Tabak; 234 saiko3p; 236 Jess Kraft; 237 Hugo Brizard - YouGoPhoto; 238 MarcAndreLeTourneux; 239 Alexey Seafarer; 240 Petr Jilek; 243 Luciano Mortula – LGM; 244 Ondrej Prosicky; 245 Volodymyr Burdiak.